DIRECT ACCESS FUTURES

A COMPLETE GUIDE TO TRADING ELECTRONICALLY

DAVID I. SILVERMAN

with a foreword by

JAMES J. MCNULTY

President and CEO
Chicago Mercantile Exchange, Inc.

John Wiley & Sons, Inc.

To Lauren

—Zeh Doh'dee, Vezeh Reh'ee—

Copyright © 2002 by David I. Silverman. All rights reserved.

Published by John Wiley & Sons, Inc., New York.
Published simultaneously in Canada.

No part of this publication may be reproduced, stored in a retrieval system, or transmitted in any form or by any means, electronic, mechanical, photocopying, recording, scanning, or otherwise, except as permitted under Sections 107 or 108 of the 1976 United States Copyright Act, without either the prior written permission of the Publisher, or authorization through payment of the appropriate per-copy fee to the Copyright Clearance Center, 222 Rosewood Drive, Danvers, MA 01923, (978) 750-8400, fax (978) 750-4744. Requests to the Publisher for permission should be addressed to the Permissions Department, John Wiley & Sons, Inc., 605 Third Avenue, New York, NY 10158-0012, (212) 850-6011, fax (212) 850-6008, E-Mail: PERMREQ@WILEY.COM

This publication is designed to provide accurate and authoritative information in regard to the subject matter covered. It is sold with the understanding that the Publisher is not engaged in rendering professional services. If professional advice or other expert assistance is required, the services of a competent professional person should be sought.

ISBN 0-471-12199-1

Printed in the United States of America.

10 9 8 7 6 5 4 3 2 1

About the Author

David I. Silverman is a long-time member of the Chicago Mercantile Exchange and a trader of financial futures for more than twenty years. He has actively traded futures throughout that time on the CME trading floor and on GLOBEX, the CME's electronic trading platform. In addition, he traded NASDAQ securities using a Level II system. Mr. Silverman also served as a member of the Board of Directors for the CME for eight years. In that capacity, he took on many important exchange positions, such as chairman or vice-chairman of various technology and product committees. During his tenure as a board member, he helped to develop order-routing and communications technology geared toward making the trading floor more efficient; chaired committees that developed a hand-held trading device and wireless headset technology; and participated in the development of GLOBEX. He also served, for four years, as Chairman of the Board of the GLOBEX Foreign Exchange Facility (known as the GFX), the CME's foreign exchange market-making division. Mr. Silverman has been an active member of the Managed Futures Association and is involved in raising money for a number of money managers. He is an accomplished public speaker and has participated on panels and delivered a number of keynote addresses at industry conferences. A 1981 graduate of the University of Chicago, he is married with four children and resides in Skokie, Illinois.

Aspire Trading Company, LLC (Aspire), is an Introducing Futures Broker specializing in online futures trading services. Founded and managed by David Silverman and Peter Mulmat, two veteran pit traders who have made successful transitions to electronic trading, the firm caters to the needs of serious traders. In its trading arcade in Chicago, Aspire offers individuals the ability to utilize state-of-the-art technology and high-speed communications to trade the futures markets with direct electronic access. Aspire also caters to the needs of customers trading from remote locations, offering clearing and educational services, as well as access to its high-speed trading networks. Aspire's mission is to lead a revolution in futures trading, whereby any individual can use their technology, training, and supervision to achieve success. The most important element of the Aspire business model is the recognition that the success of the company is tied inextricably to the ability of its customers to trade profitably. Accordingly, Aspire's central focus is to help the customer achieve the skills necessary to become a successful trader.

As the futures world moves rapidly away from its traditional structure and towards electronic trading, it is important to find a way to conquer the significant burdens technology imposes. Aside from apprehension about the upfront and ongoing costs, the biggest concerns heard from Aspire's customers are, "How do I begin?" and "Who can I trust?" That is where Aspire comes in. They can help in many different ways: by providing low cost technology solutions, competitive commission rates, and personalized training and support. If you are interested in becoming an Aspire customer, reading this book is a great way to start. But there is much more that Aspire can teach you when they work with you one on one. Every trader is different, with different needs and capabilities. Aspire will help you determine a trading plan that is best suited for you and work with you to implement it. Please feel free to contact Aspire online at *www.aspiretrading.com,* or call at (312) 638-5190.

Acknowledgments

This book would not have been written except for the initiative taken by Bill Falloon, a Senior Editor at John Wiley & Sons, who approached me after hearing a speech I delivered about electronic trading at a futures industry conference. I want to thank Bill for perceiving that a ten minute talk could be turned into an 80,000 word book and for encouraging me to take on the project. After I signed on, Bill handed me over to my editor, Claudio Campuzano. It was Claudio who, halfway into the book, listened patiently as I explained why I could not possibly meet the deadline and then helped me do what I said I could not. Finishing the book on time was definitely a joint accomplishment and I am appreciative of his good humor and understanding throughout this difficult process. Also, I owe sincere thanks to Lea Baranowski at Carlisle Communications, Ltd. and Jennifer MacDonald at John Wiley & Sons, Inc. for helping to make sure all the i's were dotted and t's were crossed.

A book like this cannot be written without the assistance of many friends and colleagues, and I am deeply grateful to everyone who helped me. I am particularly indebted to Rachna Mathur at EUREX; Karen Klitzman and Glenn Kirwin at eSpeed; Patricia Jurka and Angelo Russo at the Chicago Board of Trade; Paul Stoneham and Matthew Gibbs at the Sydney Futures Exchange; Lori Aldinger, Elizabeth Gisch, Scott Johnston, Chris Krohn, and Gail Moss at the Chicago Mercantile Exchange; Chuck Mackie, Steve Monieson, and John Bresland at Trading Technologies; Jay Wardle and Peter Hutchison at Future Dynamics; Chuck Dolce at Real Time Systems; Graham Smith at Easy Screen; Steve Lukes and Barbara Lapointe at GL Consultants,

Inc.; Thomas Theys and Ivy Chau at patsystems; and Carla Cavaletti, Tom Ascher, and Mike Domka at Interactive Brokers.

I also offer my sincere appreciation to Jack Bouroudjian, President of Commerz Futures in Chicago; Mike Persico, President, and Pace Beattie, Director of Business Development, for Tekom, Inc.; and Paul Doppelt, trader extraordinaire; all of whom were terrific interview subjects and very good sports as well. I am extremely grateful to Jim McNulty, President and CEO of the Chicago Mercantile Exchange, Inc., who, even in the midst of running the Merc—truly a 24/7 occupation—somehow found the time to meet with me, read my manuscript, and write the foreword to this book.

I would be remiss if I did not mention some close personal friends and family members who also contributed significantly to the creation of this book: Peter Mulmat, Mike Philipp, and Jeff Zaret. I also offer thanks to my in-laws, Norman and Sondra Fox, for their quiet and loving support. My parents, Marvin and Barbara Silverman, contributed to this book in a thousand ways, providing me with good advice and much-needed encouragement, as they have countless times throughout my life. My children, Jessica, Matthew, Rebecca, and Joshua deserve special praise, for allowing me to abandon them for the three months that it took to finish this project. We will spend the coming summer making up for lost time. We are all looking forward to that.

Finally, I offer my gratitude and love to my wife, Lauren. I thank her for supporting me in the writing of this book and in all my endeavors.

Contents

Foreword — XI

Introduction: The Impact of Technology on Futures Trading — 1

1 Direct Access — 9
Direct Access and Why It Is Important — 9
Defining Direct Access — 11
Types of Direct Access — 12
Direct Access and the NASDAQ Electronic Market — 12
Fragmentation and the Central Limit Order Book — 13
Trading Arcades — 15

2 Discussions about Direct Acess — 23
Direct Access and the Broker — 23
Direct Access and the Network Engineer — 32
Direct Access and the Trader — 46
The Trade Matching Engine: GLOBEX2—A Case Study — 54

3 The Case for Futures Trading — 73
Everybody Needs Insurance — 73
Hedging 101: The Basics — 74
The Speculators — 76
Organized Futures Exchanges: An Abridged History — 77
The Birth of Financial Futures Markets — 78
Open Outcry — 79
Liquidity — 79

	Illiquidity	81
	The Role of the Exchange	83

4 The Exchanges — 89

- The Chicago Mercantile Exchange, Inc. (www.cme.com) — 89
- The Chicago Board of Trade (www.cbot.com) — 101
- The Sydney Futures Exchange (www.sfe.com.au) — 106
- EUREX (www.eurexchange.com) — 110
- LIFFE (www.liffe.com) — 112
- The eSpeed Exchange (www.cx.cantor.com) — 114

5 Software — 117

- Independent Software Vendors — 117
- Questions to Ask Before Choosing a Front-End — 119
- The Vendors — 120

6 Stock Index Futures — 133

- The S&P 500 Index — 133
- The S&P 500 Futures Contract — 134
- Using S&P Futures to Hedge Portfolio Risk — 134
- Fair Value — 136
- Index Arbitrage (Program Trading) — 138
- The Tick Indicator — 140
- The NASDAQ 100 Index — 141
- The E-Mini S&P 500 and E-Mini NASDAQ Futures Contracts — 141
- So You Want to Be a Market-Maker — 146
- Trading the "Levels" — 147
- The Privileges of Membership — 148
- E-Minis: The Day Trader's Dream Come True — 150
- Tax Considerations — 153
- Single Stock Futures — 155

7 Foreign Exchange — 159

- Foreign Exchange and Electronic Trading Systems — 159
- The Interbank Market — 161
- The Foreign Exchange Dealer — 161
- Interbank Pricing Conventions — 164
- Spot and Forwards — 164
- FOREX Swaps — 167
- Macroeconomic Factors That Affect Currency Values — 168
- Trading Currencies Electronically — 169
- The GLOBEX Foreign Exchange Facility — 170

8 Treasury Futures — 173
The Need for Treasury Futures — 173
The Basics of CBOT Treasury Futures — 174
The Significance of the Cheapest to Deliver — 177
Some Real World Applications — 180
The Federal Reserve — 185

9 Becoming a Successful Trader — 191
Do You Have What It Takes to Become a Successful Trader? — 191
It's Going to Take Some Time — 192
There Are No Shortcuts — 195
Develop a Trading Methodology — 196
The Ten Commandments — 197
What Type of Methodology Should I Use? — 199
Find Your Edge — 200
Systems-Based Trading versus Discretionary Trading — 201
Understand the Value of Capturing the "Spread" — 202
Control Your Risk — 203
Understand What You Are Getting Yourself into — 206
The Investor's Bill of Rights — 208
Losses: Can't Live with 'em, Can't Live without 'em — 209
Technical Analysis — 211
These Boots Are Made for Random Walking — 212

10 Twenty-five Frequently Asked Questions about Electronic Trading — 221
1. How Much Money Do I Need in Order to Trade Actively? — 221
2. How Long Will It Take Me to Begin Making Money? — 222
3. How Much Money Can I Make? — 222
4. How Much Money Can I Lose? — 222
5. Assuming One Has to Pay His or Her Dues, How Much Money Can I Expect to Lose in the First Few Months of Online Trading? — 223
6. How Many Contracts Should I Trade? — 223
7. When I Begin, How Many Contracts Should I Trade? — 223
8. How Much Should I Try to Make on a Trade? — 224
9. What Is the Maximum Loss I Should Take on a Trade? — 224
10. What Are the Best Electronic Futures Markets to Trade? — 224
11. How Much Should I Pay in Commission? — 225
12. What Are My Connectivity Choices? — 225
13. Which Connection Is Best? — 225
14. What Is the First Step I Have to Take in Order to Get Connected? — 225
15. If I Do Decide to Trade over the Internet, Which Service Provider Should I Choose? — 226

16. Can I Connect to the Internet through a 56K Dial-Up Line or Do I Need a High-Speed Line? 226
17. What Type of PC Do I Need? 226
18. How Much Money Do I Need to Invest in the First Six Months of Online Trading? 227
19. Do I Have to Trade Full-time in Order to Become Successful? 228
20. How Do I Choose a Broker? 228
21. I Have Traded Securities Online before, but am New to the Futures Markets. What Are the Differences between Them and What Makes Futures Better? 228
22. When Can I Trade Single Stock Futures? 229
23. There Are a Number of Internet-Based Foreign Exchange Trading Sites. Why Should I Trade Currencies at the CME? 229
24. How Can I Get Started Studying Technical Analysis? 229
25. What Is the Best Technical Strategy for Me to Use to Make Money Trading Online? 230

APPENDIX A: GLOSSARY 231
APPENDIX B: INDEX COMPONENTS: S&P 500 AND NASDAQ 100 243
APPENDIX C: CONTRACT SPECIFICATIONS: S&P AND NASDAQ FUTURES CONTRACTS 249
APPENDIX D: TAX TREATMENT OF FUTURES 253
APPENDIX E: THE INVESTOR'S BILL OF RIGHTS 261
APPENDIX F: USING THE WORLD WIDE WEB 267
APPENDIX G: SUGGESTED READING 271

Index **273**

Foreword

As we enter the new millennium there is significant analysis and opinion about market rules, market structure, market "ecosystems," and market technology. Globalization, harmonized regulation, and low-cost trading technologies have given rise to new market structures and vigorous levels of competition for creating marketplaces. However, many of the early efforts have disappointed venture capitalists, investors, and, especially, early participants. The B2B marketplace mantra of "content attracts community and community creates commerce" has a hollow ring. Technology should play a large role in defining successful exchanges. However, there are many examples of new exchanges that have the self-described "killer app," a moniker for software that is faster or more reliable or provides greater functionality than others; yet, many of these exchanges still lack real traction after several years of effort. So, what does create a great marketplace?

At Chicago Mercantile Exchange, Inc., we constantly strive to answer that question. We find that the answer is not simplistic, and that it can only be found by working closely with our customers. The first thing that our customers tell us is that they want quality, consistency, and speed in pricing. They also want flawless clearing and settlement, creditworthiness, and efficient use of their capital. They want rules that are logical and dispute resolution that is even-handed. Our customers want relevant contracts that allow them to express their investment opinions or to hedge their equity, interest-rate, currency, and commodity risks in deep pools of liquidity.

Most customers do not tell us that they want either Open Outcry or electronic platforms on which to trade. They tell us to maintain, grow, and protect the liquidity that allows them to transact approximately one trillion dollars in notional volume every day at the CME. Our challenge is clear: We must innovate and constantly improve the trading platforms we provide to our clients so that we fulfill and exceed their expectations. We call this "engineering market faith."

David Silverman's book, *Direct Access Futures,* gives us insight into the early stages of the development of the electronic marketplace. Implicit in the story is the great contribution that electronic trading has made to improvements in quality, consistency, and speed of pricing. It is also true that the current methods of electronic trading and current marketplaces are only in a nascent stage. New networking developments and software functionality will improve markets tremendously in the coming years. Markets will be more global in nature and will cross the boundaries of derivatives, cash, securities, and commodity product offerings. Markets will continue to develop a backbone of professional traders that are using Automated Trading Systems to identify and execute spreads and to engage in statistical arbitrage. Exchanges will be tasked with forming efficient nodes in a networked web of transactions. Money will drive these markets, but it is the degree of faith in these markets that will drive the money.

James J. McNulty, President and CEO
Chicago Mercantile Exchange, Inc.
June 11, 2001

Introduction
THE IMPACT OF TECHNOLOGY ON FUTURES TRADING

There is a visitor's gallery overlooking the trading floor of the Chicago Mercantile Exchange (CME or Merc), the self-billed "Exchange of Ideas." Every weekday an eclectic stream of curious onlookers come to see the action. Farmers in flannel mingle with businessmen wearing Brooks Brothers. Tourists from every part the world, taken aback by the chaos on the trading floor below, conduct a symphony of puzzled whispers in a dozen different languages. The occasional trader's wife, a couple of well-groomed children in tow, balances bags from Neiman Marcus under her arms and waits impatiently for her husband to leave the trading floor to take the family to lunch. In the gift shop one can purchase a tie with a pattern of smiling pigs, in honor of the pork belly—the one CME product that the average person has heard of. These have always sold well, but lately, increasing numbers of visitors are taking home a multi-colored t-shirt with the Merc's new logo prominently displayed: "THE CME—INCORPORATING CHANGE."

The change that the t-shirt alludes to, of course, is the growth of technology. In a world where information can be reduced to bits of data that travel instantly with a keystroke, even the casual observer has to wonder whether the trading floor is superfluous. Why build a pit to accommodate a few hundred bodies in Chicago when the entire world of traders can be brought together in cyberspace? In fact, at exchanges based in Frankfurt, London, Tokyo, Seoul, Sydney, Montreal, and tens of other locations throughout the world there are no trading floors, only electronic platforms. These exchanges have grown exponentially as U.S. counterparts with trading floors have seen their market shares fall. Certainly the CME and other floor-based exchanges are aware of the profound effect technology has had on the flow of business.

Nevertheless, at first glance, one has to admit that the CME trading floor is an impressive sight. Traders in multi-colored coats jam the pits, shouting bids and offers at each other, while on the walls prices of interest rates, currencies, stock indices, and yes, pork bellies, are displayed on enormous scoreboards that can be seen from anyplace on the trading floor. In booths that surround the pit, clerks with telephones cradled between their shoulder and ear flash hand signals to traders in the pit, who relay their own signals in return. These gesticulations seem incomprehensible to outsiders, but they are the language of the trading floor, and it is clear that, somehow, these people are communicating. Surrounding the pit there is a frenzy of activity: clerks hustling paper orders into and out of the pit, traders running from one spot to another, and visitors trying desperately to avoid getting in the way. The floor is a mess, covered with trading cards, newspapers, candy wrappers, and a centerfold from a men's magazine. As one watches the show it is possible to hear the trader's frantic bids and offers through the thin glass that separates the viewing area from the pit. One's eyes are drawn to the trader's faces: the beads of sweat, grimaces, and occasional bursts of exultant glee. Most visitors are amazed at the concentration of emotion and energy—it is like experiencing a rock concert from the backstage or sitting ringside at Caesar's Palace—exhilarating and intimidating all at the same time. But first glances can be misleading, and a closer look behind the hectic facade reveals a far different picture, Many customers currently send their orders to electronic marketplaces, and the average floor trader makes less money than in years past. Most ominously, while the exchanges still do an impressive amount of business in their pits, they are worried that the cost of supporting the Open Outcry system in perpetuity will put them at a disadvantage to electronic competitors.

There was a time in the not-too-distant past when the pit was the place to be if you wanted to make it as a trader. In the 1970s and 1980s, as the word spread that there were fortunes to be made trading new financial futures products like interest rates, currencies, and stock indices, hundreds of hopeful applicants sought entry to the exchange every year. They came from a wide variety of social milieus, neighborhoods, and ethnicities. The pits became a melting pot as blue-bloods and blue-collars, high-school dropouts and college graduates, intellectuals and street toughs learned together how to create a marketplace. They had little in common but an entrepreneurial spirit and the desire to make a great deal of money quickly. Others who have written about this period have referred to them as "the new Gatsbys," but they were more than simply a class of nouveau rich. It is certainly true that they enjoyed the trappings of wealth—as every Mercedes dealer, real estate agent, and party planner on the North Shore of Chicago can tell you—but the pits produced more than just conspicuous consumers. The traders of Chicago fancied themselves as the ultimate risk-takers. Nothing intimidated them: They took on multi-national companies, banks, pension funds, even the central banks—they particularly enjoyed outsmarting them—and day after day smugly tallied up the winnings. They seemed invincible and wanted everyone to know it.

These were proud times for the Open Outcry exchanges and their members, but they were not to last. So consumed were they with their successes, the exchanges failed to respond quickly enough when, by the mid-1990s, fully electronic competitors began to capture market share from the traditional floor trading environment. The best example of this phenomenon is the case of the German exchange, known as EUREX. In 1998, the top three futures exchanges in the world were the Chicago Board Of Trade (CBOT), CME, and London International Financial Futures Exchange (LIFFE). All of these exchanges conducted their business on trading floors. EUREX (at that time called the DTB), a completely electronic exchange, was in fourth place. In 1999, however, EUREX surged ahead of the other three exchanges as major European banks and trading institutions began to make the transition to electronic futures trading. As a consequence, EUREX volume increased 68 percent over the previous year, while the CBOT, CME, and LIFFE declined 10, 11, and 39 percent respectively. Most significantly, a large part of the success EUREX experienced stemmed from achieving a 100 percent market share in the BUND contract, which until that time had been traded mostly in the pits of the LIFFE. It is no coincidence that the LIFFE decided shortly thereafter to abandon its trading floor in favor of full-time electronic trading. EUREX has continued to grow dramatically and traded more contracts in the year 2000 than the *combined* CME and CBOT pit volume. While EUREX is the most impressive example of the changing dynamic, there are many other cases that could be cited which illustrate the same point: The move toward electronic trading is inevitable. Perhaps most telling is the fact that in the last decade, of the more than one hundred exchanges that have been created world-wide, not a single exchange has chosen to conduct their business in a trading pit.

So, a year into the new millennium it is clear that electronic trading will eventually overtake and replace the trading floors. How soon this will happen is the great unknown. Everyone seems to have an opinion, but it is like trying to predict a baby's due date; you can do the math, but that baby's not coming out until he or she is good and ready. That the exchanges have moved so slowly to meet the new reality is not terribly unusual. Like most privileged classes that deny the end is coming until the peasants are outside the castle armed with pitchforks, they refuse to give up out of equal parts greed, denial, and fear of the unknown. From a practical standpoint, it will not matter much in the long-run that the exchanges have moved slowly. If they wish to see their markets grow, they will eventually capitulate, or a competitor—fully electronic, no doubt—will capture their market share.

I made my living on the trading floor for the first 15 years of my career and have spent the last few years hunched in front of a PC. Occasionally, I leave my workstation on the eleventh story of the CME and walk onto the floor before the other traders arrive. I think back and, although it has been years since I made a trade in the pit, I can still remember how exhilarating and cathartic it felt to yell out my bids and offers until my throat was raw. I realize that I miss the feel of the adrenaline rushing through my body whenever things got a bit too hairy, and the flood of relief that accompanied bailing out by passing along a bad trade to some other trader with a bigger account and who knows what else bigger than mine. I recall certain events with remarkable clarity—how the pencil felt as I scratched the particulars of a trade onto a trading card; feeling my shirt stick to my body in a fast market, soaked with sweat as if I had been running in the hot sun; or fighting the urge to count my booty after making a winning trade. In the early morning quiet, it seems like no time has passed—that I can still fit into a size 40 trading jacket, that the picture on my CME identification card shows a young trader without a gray hair, and that today, if I wanted, I could take my old spot in the pit and teach the new guys how to really throw some numbers around. Then, invariably, I stretch and feel the tightness in my lower back. I remember what it was like to stand in a crowded pit. The years of physically exerting myself—screaming for trades and elbowing for position—have done their damage and, like an aging athlete, I know my body has only begun to punish me for the excesses of my youth.

Although I have chosen to leave the floor, and an increasing number of my colleagues have as well, in some ways it will be sad to see the pits fade away. There is a sense of history here; it is in the floorboards upon which restless traders stand their ground and in the rafters where the echo of countless bids and offers is absorbed. How many fortunes have been made and lost in these pits? In them, the traders of Chicago created a grand expression of capitalism, original and full of vitality for most of its life. That the glory years are over for the trading floor does not diminish the importance of what transpired on it. No matter how ignominious its end, the world will never be the same for what was created here. No one knows when the last shout will be heard on the trading floor, but even those who are most recalcitrant are planning for a future in which the sound of a trade is no louder than the click of a mouse.

In the coming years, perhaps the most significant change to the marketplace will be the increase in the number of traders who can participate as equals. On the trading floor, space limitations necessitate that some traders—those that stand close to the order flow—are more equal than others. They have first crack at the most lucrative orders and generally make the most money. An electronic environment, however, is one in which every market participant "stands next to the order flow." This new democratic approach will lead to tremendous opportunities for everyone in the marketplace.

Therein lies the purpose behind this book. It is certainly correct to suggest that, with the inevitable changes ahead, there will be great opportunities for online traders, but one thing will remain the same: those who are unprepared will continue to lose money to the market professionals. In fact, because the technology is so powerful, they may be at an even greater disadvantage to the professionals than before. For those who are serious about capitalizing on the opportunities ahead, it will be necessary to learn about market dynamics, technology, and strategy in the online world. That is what you will find here. After you read this book you will be ready to begin the task of becoming a successful trader and have a reasonable chance of achieving your goals.

The book is divided into ten chapters. The first chapters are devoted to an analysis of "Direct Access": what it is and why it is important. In addition to an overview and examination of the key issues associated with direct access, you will learn what three market professionals—a network technician, a broker for a large Futures Commission Merchant (FCM), and a former floor trader who has successfully made the transition to electronic trading—have to say about the subject. Their knowledge and insights are relevant to both experienced traders and novices. Finally, you will see an electronic trading platform—the CME's GLOBEX 2—dissected into its various components. This will enable you to better understand the functionality that can be expected from an electronic trading system. In Chapter 3, we will take a look at "The Case for Futures Trading." In it, we will examine some basic information about the futures markets: how they work, who uses them, and how. We will also examine the vital role that the exchanges play and how the recent phenomenon of "demutualization," in which the exchanges seek to transform themselves from not-for-profit to for-profit businesses, will ultimately lead to their transition away from the Open Outcry system. Chapter 4 presents an overview of the various major exchanges, both domestic and foreign, including their products, hours of operation, and pricing information. Chapter 5 covers software choices. The first part of the chapter gives an overview of the major Independent Software Vendors (ISVs) who provide front-end trading systems (simply put, the software the trader uses to see market prices and enter trades into the market). In the latter part of the chapter we take a look at some of the major technical analysis software packages. Chapters 6, 7, and 8 present an analysis of three major trading products: the e-mini S&P 500 and NASDAQ 100 contracts, currency futures contracts, and long-term interest rate futures. In Chapter 9, you will find a number of strategies and approaches that will help you trade the market like a professional. In Chapter 10, the final chapter, we will review the key topics covered in the previous chapters through a list of 25 questions frequently asked about electronic trading. The appendices include an extensive glossary of industry terms; a list of the components of the S&P 500 and NASDAQ 100, as well as contract specifications for the futures products on those indices; the text of the tax law covering futures transactions; the "Investor's Bill of Rights"; a list of useful websites; and a suggested reading list of important books about trading and the futures markets.

I have lived in the futures world for my entire professional career. I have traded bull, bear, and sideways markets; currencies, bonds, and stocks; slow markets, fast markets, and markets so dangerous that they should have come with a warning label. (The Surgeon General has determined that this market can cause large losses. Possible side effects can include headaches, weeping, and thumb-sucking.) It is not a perfect world, and sometimes life here is difficult. However, it is a world filled with surprises; no two days are exactly alike. I look forward to showing you around my world. Let's get started.

1
DIRECT ACCESS

DIRECT ACCESS AND WHY IT IS IMPORTANT

When I started trading on the floor of the Chicago Mercantile Exchange (CME) in the early 1980s, I was advised by the established traders that the exchange required all new members to stand in the center of the trading pit. I was told that it was the best place for me to work on my trading techniques and the best vantage point from which to watch the successful traders operate. This, of course, was nonsense—the exchange had no such rules and I was free to stand wherever I liked—but I did what I was told, hoping that it would ease my acceptance into the group. Still, I had an uneasy sort of feeling, like a victim on a bad episode of *The Twilight Zone.* Had I somehow been transported back in time to high school where the seniors get to push around the new students? I wasn't entirely sure of the agenda, but I was certain that those "advising" me on this matter had little interest in my welfare or career advancement. I was right about that.

Trading in the center of the pit was like being told by your coworkers, after the entire box of donuts is eaten, to "help yourself to the holes." After only a few days of screaming bids and offers at the traders on the higher steps, with little to show for it but a raw throat, sweat-stained trading jacket, and a handful of unprofitable trades, I was almost ready to call it quits. I saw how jealously the top-step traders allocated the most lucrative customer orders for themselves and knew that if I didn't figure out a way to get up there, my career as a trader would primarily be notable for its brevity. Whatever camaraderie I felt with the traders who were now ignoring me or desire I had to please them had been replaced by feelings best left unexplored. I was exhausted, my throat felt like I had swallowed a pound of broken glass, and most of all I was hungry. I wanted my donuts.

In fairness, there was a practical reason for this caste-like discrimination: It was the veterans who were best equipped to handle the crush of orders executed on the top steps of the pit. They were better-capitalized and knew how to provide the liquidity that the customers demanded. Granting a novice trader access might actually infringe on the ability of the more experienced traders to accommodate the order flow and create liquidity. While the needs of the marketplace were certainly met by such an approach, it was hardly altruism that served as the foundation of the pecking order. Generally, the top-step traders cared little for the welfare of the customer. What they did care about, desperately, was their own physical proximity to the orders those customers were sending to the pit. They knew that the closer one was situated to order flow, the closer one was to the best trading opportunities.

Meanwhile, in the center of the pit, I tried to make the best of a bad situation. Occasionally, a few trades filtered down, allowing me to join the fray with the more active traders on the steps above. Each day I made a few more trades than the day before, and through hard work, patience, and careful risk management I began to see the equity in my account grow and with it my self-confidence. One day, I came to work an hour early and planted myself on the top step of the pit and waited for the rest of the traders to show up. As you can imagine, they had no intention of welcoming me into their neighborhood. In fact, they did everything they could to force me out, from calling me names to elbowing me in the head, ribs, and kidneys. Whenever it got busy and we were pressed tightly together, one trader took to "accidentally" writing on my collar with his pen, which happened to be an indelible laundry marker. I took it all: the nasty language, the sharp elbows, and even the ruined shirts, because from my new top-step location I began to make some real money. I figured my ego was strong enough to take the insults, my elbows were as sharp as anyone else's, and I could afford to buy all the new shirts I needed. At some point, the other traders stopped trying to intimidate me. To be sure, each one had followed a similar course to their respective positions in the pit and knew that I had earned the right to my upward mobility. As my volume and stature in the pit grew, I realized I could have whole boxes of donuts if I wanted and that I'd never be left with the holes again. Before long, I was counseling new traders on the importance of learning how to trade in the center of the pit, although as some moved up over time I spared them the verbal and physical abuse. The point of this story is at the heart of what you will learn throughout this book: *A trader will always be more successful with direct access to order flow.*

DEFINING DIRECT ACCESS

It's important to define direct access and it is interesting to note that the definition is applicable for both electronic and non-electronic markets:

> Direct access is the ability to participate on the current bid and offer with other market participants and to be able to see and have access to the order book and other useful market information.

Probably the most profound change that technology will bring to the futures market is that direct electronic access will soon be available (and in some markets is already available) to anyone who is willing to pursue it. The cost of achieving direct electronic access is far lower than the cost of buying a seat on an exchange and then paying dues while trying to establish a physical presence in the pit. Also, direct electronic access is *better* direct access; it is available to every market participant from the instant that he or she logs on to trade. In the pit, there is always a "best spot"; it's the area in which the majority of the customer orders are executed and, of course, can only accommodate a finite number of traders. In an electronic marketplace, by contrast, there are an infinite number of best spots. More specifically, if we define every PC with direct access as a spot, we can say that every market participant is in an equally good spot. This is an extraordinary leap forward from the days when I fought for a top-step position on the floor of the CME. It means—for the first time in the history of the futures exchanges—that everyone can profit from direct access to the order flow, without ever having to throw an elbow into another trader's head.

TYPES OF DIRECT ACCESS

Direct access, broadly defined, encompasses different types of technology and connectivity. One can, for instance, obtain direct access by connecting directly to the exchanges and their electronic platforms. This type of connectivity is indispensible to those who want to trade the market actively. It is common to find this type of access in professional trading rooms. On the other end of the spectrum, one can log onto the web site of a broker that is itself connected to various exchanges. While this is a form of direct access, it is actually direct access to *the broker rather than to the marketplace.* This is an important distinction. By allowing the broker to act as an intermediary, the electronic trading process slows down. It is quite similar to the difference currently experienced in the open outcry system between sending an order directly to the trading floor and "flashing" it into the pit—traditionally the quickest way for a customer to enter an order into the market—or calling a broker in an office who records the order and then calls the trading floor to have it flashed into the pit. The basic principle is the same: intermediaries and intermediate steps *always* slow the trading process and ultimately add to the cost of making a trade.

That does not necessarily mean the faster approach is the only acceptable method or even that it is better for every trader. In order to know which approach is best suited for you, it is necessary to determine what type of electronic trader you hope to become, what your expectations are, and how much of a financial commitment you are willing to make to technology. Generally the more active one is, in terms of numbers of transactions and total volume, the more critical it becomes to trade using the fastest, most direct access to the market. To the extent that one is not so active, it may be less imperative to seek out and pay a premium for the most direct access.

DIRECT ACCESS AND THE NASDAQ ELECTRONIC MARKET

Direct access in the securities industry began, for all practical purposes, in January of 1997, when far-reaching rule changes facilitating direct access to the National Association of Securities Dealers (NASDAQ) electronic marketplace went into effect. Prior to that time only broker-dealers and marketmakers could obtain direct access. The rule changes, however, allowed non-market professionals to have direct access to what is known as NASDAQ Level II information for the first time. Immediately, individuals and groups of electronic day traders in trading rooms across the country took advantage of this newfound direct access and began to trade actively, placing their bids and offers on the inside market, just like the broker-dealers and marketmakers.

Initially, the professionals decried this turn of events. Like the established traders I encountered who encouraged me to stay in the center of the pit, broker-dealers and marketmakers were reluctant to share the order flow with these new market participants. But the rule-makers who mandated the changes, and the technology providers who created powerful electronic trade matching systems, called electronic communication networks (ECNs), ensured that the effect on the market was immediate and profound.

While it is not the intention of this book to examine the development of direct access in the securities market in great detail, there are some lessons futures industry professionals can learn from the growth of electronic securities trading. In order to appreciate those lessons it is important to understand the technological infrastructure that supports electronic securities trading and how the various market participants communicate with each other when they want to make a trade.

FRAGMENTATION AND THE CENTRAL LIMIT ORDER BOOK

Transactions in stocks listed on the NASDAQ can take place in a number of different ways. Orders can be sent to:

- the Small Order Execution System (better known as SOES)
- the SELECTNET system, which allows traders to specifically "preference" or choose a particular marketmaker on the Level II system
- an ECN, such as *Island, Instinet,* or *Redibook*

In order to understand how this works, imagine that every message (e.g., a bid, offer, rejection, or an actual transaction) sent on any of these systems is a car on the entrance ramp to a highway trying to merge. If every car/message tries to enter in an orderly way, following a common set of rules, traffic will flow smoothly. But if cars try to enter wherever, whenever, and however their drivers wish, with little regard for the other cars on the highway, there will be chaos; or, in the parlance of the marketplace, *fragmentation*. Each of the millions of SOES, SELECTNET, and ECN messages sent during a trading session is like one of those cars trying to speed ahead of the others onto the execution highway, making NASDAQ trading as stressful an experience as being caught in an endless rush hour traffic jam. Because it lacks a *Central Limit Order Book* (CLOB) or, to use our analogy, a place where all cars would head on a single entrance ramp and from which they would return when they wanted to come home, trading is inefficient, unstable, and frequently breaks down. Joey Anuff, the author of *Dumb Money: Adventures of a Day Trader* (McGraw Hill, 2000, p. 105), tells a story about what it is like to be caught in a NASDAQ traffic jam:

> I learned that the mechanism of a trade was almost indecipherably complex . . . I could hit a posted bid or offer and pay the spread if I wanted to but I could also post bids or offers of my own and let other traders hit me. I could choose between routing my orders to a private ECN like Archipelago or Island, or I could send it to the NASDAQ marketmakers via SelectNet. It was all a matter of timing. And hand coordination. And something else. Some magic. Some mojo. Time and again my trades were tangled in execution snafus.
>
> One morning I noticed a new issue had debuted. APLN, priced at 14, opened somewhere around 15, dropped to 14 and made a lame run to a hair above 16. Getting in was easy. Instinet (an ECN) showed 1,000 at 16 1/8 and I took the whole offer. It looked like a good trade and the price moved right up. For a second. Then Instinet came back with 1,000 more shares. At 16 1/16. What the hell was I doing here? I made to get out. As I plugged in a sell order at 16, I recalled seeing a warning come over the wire that SELECTNET was down. Which was a problem because SELECTNET was my best way of selling all 1,000 shares to the marketmaker on the bid at 16. Maybe SELECTNET is back up, I reassured myself as I pulled the trigger and entered a world of pain.
>
> The sell order wasn't rejected. It just didn't go anywhere. Somewhere, perhaps on my broker's network, maybe on a NASDAQ server somewhere in New York, my order sat stalled in a frozen queue as I watched APLN dip below 16. My cancel orders went unaccepted. I tried to sell again, but since my last order was still pending the automatic order processing system assumed I was trying to go short. And you can't short an IPO. My new sell orders were rejected. Now APLN was at 15 7/8. Online brokers have backup customer service representatives sitting by the phone for just such emergencies

as this. My broker picked up the phone on the second ring. I tried to explain the situation. He promised to check into it and call me back. Five minutes later APLN was at 15 5/8. I called back and got somebody new and tried to explain again. They transferred me to the trading desk and I got disconnected. APLN made a leisurely retreat to 15 3/8. Back on the phone a third person listened to my story and asked me whether I wanted her to sell my position for me. "Well, I'd rather just have the 1,000 shares in my account so I can sell them myself," I answered. I was still watching the screen. Hello, 15 1/8. But "barring that," I added smoothly, "yes, I suppose you should sell, sell, SELL!"

Sold at 15 1/16. The price tag for this three-minute fiasco: $1,053 plus commissions.

There are many reasons why NASDAQ does not utilize a CLOB and probably never will. That part of the subject is not particularly relevant to our study. The key point for us to recognize is that *every electronic futures system operates with a CLOB*. As a result, while no electronic system is foolproof, futures platforms tend to be quite stable and far more efficient than that of NASDAQ. Most importantly, trading in a system with a CLOB ensures that all market participants are treated as equals. It is upon this foundation that we can begin to build a formula for trading futures successfully.

TRADING ARCADES

The term "trading arcade" came into vogue in the late 1990s when the Marché A Terme d'Instruments Financiers (MATIF) in Paris and The London International Financial Futures Exchange (LIFFE) in London abandoned their respective trading floors, after which former pit traders began to trade electronically from remote trading room locations. I've never liked the term; it conjures up visions of dark, noisy rooms in which money is frittered away on video games, fast food, and garishly colored stuffed animals. Nonetheless, it has definitely taken hold over more sedate phrases such as "dealing room," or simply "trading room." I suppose my objection is mainly semantic, but to the extent the perception is formed that *trading* in an arcade is comparable in any way to *playing* in an arcade, the term creates a false impression. Trading in an arcade (with my reservations on the record I am willing to use the phrase) is serious business. Anyone who confuses the purpose of the trading arcade is likely to lose a great deal of money and head home without even a stuffed animal to show for the trouble.

That settled, we can now examine what constitutes a trading arcade and why arcades are the logical and necessary successors to the trading floors of the open outcry exchanges.

What Is a Trading Arcade?

An arcade is basically any physical location at which traders gather together to electronically trade. In fact, to the degree that they have introduced limited forms of electronic trading to the trading floors, it can be said that the major U.S. exchanges are themselves arcades. An arcade may have a hundred or more traders in it, or as few as two. The traders may trade for a proprietary "house" account in which all in the house share in the trading results, or they may be individual customers, each trading his or her personal capital and responsible for his or her own profits, losses, and business expenses. Demographically, arcades are melting pots, mixing professional traders with novices, mid-career changers with those fresh from a university, and both men and women (at this time there are still far more men, but the disparity in numbers is diminishing). Many different products may be traded in an arcade. A single screen may accommodate electronic markets in NASDAQ securities, stock index futures, cash bonds, and scores of other products from exchanges all over the world. The traders may have access to analytical information such as charts and news services, software that allows them to test trading ideas, and Internet connectivity. There is also likely to be a squawk box service that broadcasts live prices from various open outcry markets. Finally, no trading arcade is complete without a television monitor hanging from the rafters tuned to CNBC.

Why did arcades evolve and why have they become so popular? There are five primary reasons: *cost savings, network construction, technical support, educational support,* and *sense of community.*

Cost Savings

The reason traders initially came together in an arcade setting was to save money on creating the infrastructure to support active trading. While there are many substantial costs involved in setting up an arcade, there are also significant economies of scale that can be appreciated as the number of traders who utilize the infrastructure grows. In fact, incremental costs diminish sharply with growth. Interestingly, this is in stark contrast to the exchange's trading floor environment where adding more traders entails *higher* incremental costs and where it is physically impossible to add more than a finite number of traders. So while the floor can never grow beyond a certain threshold, the arcade environment can support exponential growth.

Let's consider how this works in real life. Connecting directly to the exchange entails thousands of dollars in upfront costs and thousands more in ongoing expenses such as rent, high-speed phone lines, license fees to the exchange and software vendors, and maintenance of the infrastructure. In an arcade, however, the traders appreciate the benefits of economies of scale. High-speed phone lines, for example, cost $2,000/month to support either one trader or fifty. With fifty, however, the cost per trader is a manageable $40/month, 50 times less than the individual is forced to pay for the same level of service. As if the economies of scale argument weren't a compelling enough reason for the individual to run to the nearest trading room, arcade traders are also able to get the benefit of receiving discounts because they buy in volume. This range of benefits extends from purchasing equipment to negotiating preferential commission rates using the arcade's aggregate volume as leverage. Every arcade is different and the scope of the benefits varies, but one thing is indisputable: By joining together to share expenses, traders in an arcade can enjoy premium services at the lowest possible cost.

Network Construction

In an arcade, trading workstations are joined together in a network that allows many PCs to be connected to the markets through the efficient use of hardware, software, and high-speed phone lines. A properly constructed network allows the trader to appreciate the economies of scale previously mentioned. There is an additional important consideration: How a network is constructed (i.e., whether it is fast, robust, and easy to maintain) will, in many cases, determine the viability of the arcade and the success of its traders. If care is taken in building the network, users can achieve a significant competitive advantage against those who are trading in less-robust networks. In building the Aspire Trading Company arcade, for example, I established, as a minimum standard, that the networks could be no slower than the fastest network allowed for by the laws of physics and rules of the exchanges. The upfront costs to accomplish this were substantial, but were justified because the payoff was immediate. With the first transaction, everyone on our network recognized that no other trader in the world was going to beat us to a trade. Trading with that sort of an advantage is one of the few ways possible to gain a substantive edge over the competition in an electronic environment.

Technical Support

One of the reasons using technology is so frustrating is that PCs, networks, phone lines, and software always seem to be breaking down or not working at optimal levels. This is true in any environment—consider the person who waits on hold for an hour with a bank or airline only to be told to call back later "because the computers are down"—but especially so when it comes to electronic trading systems. In fairness, it must be noted that trading networks are enormously complicated pieces of technology and that they need to support vast amounts of information being sent at very high speeds. Furthermore, because the electronic trading industry is so new, much of the technology being used is "buggy" and prone to crashing.

Simply put, these systems break and you need to know how to fix them when they do. While individuals can take responsibility for technical maintenance and support of their own PC and connectivity to the market, most people don't have the necessary interest, experience, or diagnostic skills to do so effectively. In most cases, they will end up contacting someone for help. This approach is highly inefficient, costly, and likely to eventually result in situations where the trader is without access to the market in the wake of a system crash. In an arcade, however, the individual is freed from these concerns. Typically there is a technical specialist on site or on call, and that person is responsible for troubleshooting and problem solving. In the Aspire arcade, we take an additional precaution by having an entire back-up system available that we can run in place of our main system in the event of an emergency. We hope we will never have to use it, but if it ever becomes necessary, our customers can continue trading seamlessly while the technicians concentrate on fixing the problem.

Educational Support

Most people who come to trade in an arcade are completely intimidated by the technology. In the same way that a novice pit trader's palms will sweat and pulse will quicken before he screams "BUY THEM!" for the very first time, so too will the new electronic trader feel insecure with a mouse in his hand and $50,000 of margin money burning a hole in his pocket. In the best arcades, the arcade sponsor (in the securities world, a broker-dealer; in the futures world, a Futures Commission Merchant (FCM) or an Introducing Broker) addresses this problem by providing the customer with intensive training that may last anywhere from a few days to a couple of months. The training includes, at a minimum, a review of the trading software that the arcade uses and can be as involved as providing one-on-one mentoring until the customer feels comfortable enough to trade without supervision. Theoretically, at that point, the trader is ready to start making money.

But do they make money? It is argued by securities industry critics that most day traders are not successful, although no definitive studies have been performed. There is so much controversy about this subject on the equities side that the Securities and Exchange Commission and Congress have commenced an investigation into the industry. Two great concerns that have emerged out of the investigation are:

1. Some broker-dealers may not be supervising their clients properly
2. Some broker-dealers may be encouraging their clients to churn their accounts

In one instance, state regulators in Massachusetts shut down a broker-dealer when they determined that 67 out of its 68 customers lost money day trading.

In the securities industry, one firm that takes the responsibility of educating its customers prior to allowing the transaction of a single share of stock is *Broadway Trading,* a broker-dealer with whom I opened an account a few years ago. *Broadway* operates a number of trading rooms on the East Coast and also supports many customers in trading from remote locations, connecting to the firm's trading network through dial-up phone lines. The firm's customers go through a rigorous training program and the company's principals carefully supervise them during the program and thereafter. One thing I learned after attending the training program is that going through it will not guarantee success, but not going through it will almost certainly guarantee failure.

The firm is bold enough to publish the profitability statistics for its hundreds of active day traders on its web site (*www.broadwaytrading.com*). The data reflects that more than half of *Broadway's* traders are profitable. In an industry in which the percentage of failure is said to be in excess of 85 percent, its commitment to education stands out. As the electronic futures industry evolves it is likely that the firms and traders who will be most successful will be those that focus on training and education. In my firm's arcade, we are mindful of that and spend a substantial amount of time, effort, and money mentoring traders one-on-one, particularly through the very trying first months of trading. We recognize that if our customers do not develop the skills to trade profitably, they will not remain customers for long.

A Sense of Community

At the height of the bull market in equities, the financial news networks were inundated with television commercials from brokers who touted the benefits of online day trading. In one, a young man who hates his "real" job is trading online and is ecstatic to see that a stock he owns is shooting off of the chart. He pumps his fist, kisses his astonished secretary full on the mouth, and storms into his boss's office and tells him to go to hell. By the time he has returned to his desk, however, the stock has plummeted and as the camera fades to black we hear the unfortunate young man tentatively asking for his job back. In a different version of this genre of commercial, a young man enters his buddy's apartment which is strewn with pizza boxes and coffee cups. The buddy, who is unshaven and looking as if he hasn't showered in days, is sitting in front of a PC in his bathrobe and boxer shorts. He is mesmerized by the blinking lights of his trading screen and recites in a monotone: "I'm up a teeny, up a half, just made $62 bucks, I'm da' man, oh no, my stock is tanking, oh no, I'm ruined! No wait its coming back, I'm okay."

I'll leave it to others to debate whether these types of marketing campaigns depict an image that encourages profitable trading activity (although one has to admit the ads are very funny). They do underscore, however, that environment can affect the trader's ability to function at a high level. A clean, comfortable workspace where one can concentrate solely on trading—without outside distractions like a "real job"—

is a prerequisite for anyone who wants to make a living as a trader. Occasionally one hears stories about individuals who trade successfully from their home offices or money managers who trade from their yachts somewhere off the coast of Puerto Wherever, but it is extremely hard to trade on your own. Of course, those individuals who do manage this feat enjoy the benefit of being able to trade in a bathrobe.

Traders tend to do better when they interact with other traders; they are able to share ideas, learn from each other's successes and failures, and provide each other with intellectual and emotional support. It is extremely difficult to achieve that level of interaction and bonhomie when alone, even if one is connected to the outside world by phone, fax, and modem. The camaraderie that develops in an active arcade is perhaps the greatest benefit of all. It helps the trader sit in front of a monitor for hours and hours and concentrate completely on trading. On occasion you may find your neighbor in the arcade talking about last night's game or a hot date, but when someone in the room yells, "Shut up you morons, I'm trying to trade," the offensive conversation will cease. When you're in a trading room, you focus so intently on the market and the action on your monitor that when it comes time to act you are prepared, as a friend who trades NASDAQ stocks at an arcade in New York says, "to eat the glass." In an arcade full of successful traders, that is exactly what you will see: a hundred hungry glass-eaters who chew, swallow, and digest the shards and come back ready for seconds.

If You Can't Trade in an Arcade

You should make every effort to trade in a professional setting with other traders (even if it's just with one or two friends), but sometimes it is not possible to do so. One major impediment is that because the electronic futures industry is so new there are relatively few arcades where one can trade. As the futures industry grows in the next few years this will change. It is likely that, at the very least, there will be futures trading arcades in every city in which one can now find an arcade for securities trading. However, if you currently find it impossible to share an office with colleagues or find an arcade, don't get overly discouraged. It is possible to successfully trade by yourself, but you should be aware that the odds of succeeding are far less favorable and, as previously stated, the costs will be higher.

2
DISCUSSIONS ABOUT DIRECT ACCESS

DIRECT ACCESS AND THE BROKER

As direct electronic access to the markets becomes the standard, there is a widespread misconception as to the role that will be played by the FCM. Some have predicted that when the customer can connect to the markets directly, there will be no need for a broker to act as an intermediary. To some extent this is true; various functions that the broker traditionally performed will no longer be necessary. The FCM, however, will have other important responsibilities for which customers will be willing to pay fees. Jack Bouroudjian, President of Commerz Futures (*www.commerzfutures.com*) in Chicago, is an example of a broker who is changing with the times. Jack came onto the trading floor of the CME at about the same time the exchange introduced stock index futures. As volume in the pit exploded, he built his business by providing value-added services that customers used to obtain an edge. With the advent of the e-minis, he evolved into a broker capable of catering to the needs of customers who required pit services and those who trade electronically. Jack no longer brokers trades next to the pit every day; he currently concentrates on building his firm's existing customer base as well as mining for new electronic futures traders from all over the world. His insights provide a view as to how the FCM community perceives the growth of direct access and the opportunities that will result from it.

Interview with Jack Bouroudjian, President, Commerz Futures (Chicago)

QUESTION: *Jack, could you please tell us something about your background?*

JACK: I spent about 18 years on the trading floor before going "upstairs"; mostly in the equity quadrant, next to or in the S&P 500 pit. I was there basically from Day One. I traded for three years in the pit; I've handled customer orders, brokered customer orders, and I've sold customer orders, so I've handled all facets of the floor and seen the evolution of the floor as far as it goes. I am currently President of Commerz Futures, a wholly owned subsidiary of Commerzbank, about the twentieth-largest bank in the world.

QUESTION: *As a CME director for many years, you've been intimately involved in virtually every important initiative regarding the CME's stock index futures contracts; in particular you were one of the first proponents of the exchange's hugely successful e-mini contracts. Can you tell us why you think the e-minis have become such a success and what the future holds for them?*

JACK: The e-minis were a perfect business model. Right when we saw the power of the Internet the business model started to change, and all of a sudden the model became "cannibalize thyself" because you didn't want someone to do it for you. So I think what's happened is we at the CME followed that plan and added a few nice variations; we launched products that complemented what we had on the floor. We created a synthesis. If something that is pure screen-based is the antithesis of something purely outcry, we created a brand new type of electronic product whose liquidity is based on volume and open interest generated in the pit; an interesting concept which I understood from the beginning. I never believed the pit "open interest" would migrate totally to the electronic platform because it was the complementary nature of the products that made it work. We gave the end user exactly what he wanted—full control of the order while still having a liquidity base in the pit that is unparalleled. Until we can figure out a way to migrate that liquidity to the screen—and we may never fully get there—you will continue with this bifurcated market. But that's not necessarily a bad thing. There is an institutional market and a retail market. A lot of that is because the needs are different. A retail customer, who is used to having an account at E-Schwab or CSFB Direct, is used to instantaneous fills and access to information. The institutional user, however, wants something completely different; they want a read on the market. It's much more three-dimensional. In the pit you get a feel for the nuances of the marketplace that cannot be found on the screen, which is a two-dimensional experience.

QUESTION: *In addition to helping design and market the e-minis, you have also been involved in the politics of regulation of these contracts. One issue on which you have expressed a strong opinion is keeping margins on stock indexes—or as we call them at the exchange, performance bonds—at reasonable levels, so as not to choke off liquidity. Can you touch on some of the issues the exchange faces with respect to margins as it fights a constant battle with the regulators about this matter?*

JACK: It's a three-way problem: The Fed versus New York versus Chicago. Obviously there is opposition in New York to increasing leverage in the futures markets because their products, which are margined at 50 percent, have to compete against stock index futures which can be traded with only a 5 or 6 percent margin. That is inherently one of the problems that the futures exchanges face. On top of that, the regulators in Washington, still sensitive to criticism that the futures markets caused the 1987 crash, are very aware that there is an incestuous relationship between derivatives, options, and cash, and how important it is to monitor all of those. Because of that dynamic it seems as if the stock index markets are looked at differently than Eurodollars, Corn, or Pork Bellies. The ironic thing is that even with the additional leverage available, stock index futures pose no systemic risk. In fact, because they are settled the next day, or *marked to the market,* they are probably less risky than stock transactions, notwithstanding the additional margins that stock traders must put up. The mark to the market facet is the great equalizer. That is essentially the problem as we launch stock futures—how do you float a mark to the market solution when the securities world works on a T+2 settlement basis?

QUESTION: *Another issue on which you have been in the forefront is circuit breakers. There are some who argue that circuit breakers are unnecessary and even counter-productive, that in a falling market trading halts don't attract buyers—value attracts buyers. Others say that circuit breakers protect clearing firms and customers from disaster when the market is in free-fall mode. How do you feel about circuit breakers as a market participant and president of a clearing firm?*

JACK: I'll give you two answers. My philosophic answer is that I hate limits of any kind. They are targets and usually attract the market. But having said that, as president of Commerz Futures I must say that I am in favor of "speed-bumps." I don't like market halts because they are counter-productive. Speed-bumps, on the other hand, give people a time to breathe, a time to re-evaluate and, many times, a chance to look for value. People sometimes forget this. Obviously, speed-bumps are only needed on the way down. I say that because the psychology of a breaking market is much different than that of a rallying market. There's euphoria when the market is going up; there's panic when the market is going down. Therefore, I think that speed-bumps are essential because they allow the investor a chance to look for that value which in many cases he cannot see because of the panic around him. Every good trader will be able to find that gem of opportunity, if it's there, as long as he is not distracted by the noise all around. So for that reason, I find speed-bumps an important component of the market structure.

QUESTION: *With the implementation of the Commodity Futures Modernization Act, or CFMA, traders in the United States will soon be able to trade Single Stock Futures and Narrow-Based Indices. How viable are these products? Is there a chance that if SSFs are successful that they will cannibalize business from existing stock index futures contracts?*

JACK: The launch of Single Stock Futures is significant. The detractors, the biggest opponents, are making the exact same arguments they made when the CME started trading stock index futures in 1982. I really find that ironic. The marketplace will ultimately decide, as it always does, how and where these instruments will be traded. There are obviously a couple of problems. Right now, the way the law stands it is very cumbersome to trade SSF, taking away the benefits of futures. I'll give you an example. If you're long IBM stock and sell an IBM future against it, it is treated as a constructive sale subject to tax. Certain things, like that, need to be changed. But these rules, and ambiguities that came out of the CFMA, arose out of the infighting between the various exchanges in Chicago. Now, however, there is a spirit of cooperation because of the new for-profit venture between the CME, Chicago Board Options Exchange, and Chicago Board of Trade to trade single stock futures. I think it's significant because, for the first time, it begins to coordinate efforts toward a common goal. We should see more than just a coordination of business activities. We should also see a coordination of political goals and efforts. In fact, that might be the most significant element of the venture between the three exchanges; working together rather than trying to undermine each other at every turn.

QUESTION: *The concept of SSF, as envisioned under the CFMA, raises the issue again of regulation. The SEC and CFTC are granted a sort of joint custody of the products and are instructed to solve outstanding issues like margins, tax treatment, and suitability concerns. Do you think the SEC and CFTC will be able to work together effectively? Does their joint involvement as regulators raise or diminish the chance that the SSFs will become successful products?*

JACK: If the CFTC and SEC work together to provide regulatory relief we will see an explosion in the futures industry. Part of the reason selling SSF may be problematic is because Series 3 Registered Brokers have only up until now sold to customers in the futures industry. Without an expanded distribution network SSF will have a difficult time becoming established. You need to establish that distribution network so that you are out there in the forefront . . . so that you're mainstream. In many ways, futures aren't mainstream. They are not sold to many people who could use them and need them. Especially over the course of the last year, where so much wealth was created and then destroyed, the idea of the "hedge" has become acceptable once again. It's very important for people to become educated and use the right risk management tools because they are right at their fingertips. Unfortunately, in some cases they may have already asked their brokers to provide futures services, but have not been able to get what they wanted because of legislative problems surrounding derivatives. But, if you get the CFTC and SEC working together toward a common goal, then the sky is the limit.

QUESTION: *Let's move on to the topic of electronic trading. Your firm supports customers who want to do their business on the trading floors as well as those who are completely electronic. What are the special challenges that you face in managing the respective needs of this diverse customer base?*

JACK: We've realized that there are different types of problems and risks associated with electronic platforms: there are risks that the system will fail, that a false position will be sent, and many other types of risks that we never expected to have when we first dreamt about these systems ten years ago. We always envisioned that being able to trade on our desktop would be as dependable as having an EXCEL spreadsheet on our PC and you would no more think that the system would crash than you would think the spreadsheet would go down on your screen. We need to get to the point where bandwidth and technology get us where we are completely comfortable and have the most dependable, web-based systems. I think we're getting there and, now, are busy identifying our customers' needs. Certain customers just want access to find out where the market is. In many cases, they want a front-end just to be able to monitor their positions and they are willing to pay for that. Other customers want to be able to actively trade the market. We, for instance, provide connectivity to the London International Financial Futures Exchange (LIFFE), EUREX, CBOT®, and the CME. We see trades in all of those markets across the board. We are trying to provide everyone with the tools they need, but it's dynamic and changing all the time.

QUESTION: *For many years, as one of the top brokers on the trading floor of the CME, you facilitated and executed an enormous amount of stock index business in the pits for your customers. Are you finding that they are continuing to do all their business in the pits? To the extent that they have migrated to electronic trading at all, is your role to lead them in that direction or simply to help them once they have made the decision to switch?*

JACK: My role is to provide the customer with whatever he (or she) wants. Right now, I am supporting two platforms because, quite frankly, I have customers utilizing both platforms. It's not for me to lead the customer to anything. It's for me to be able to provide the customer with what he (or she) wants. In fact, that's always been my approach. Anything that is within the law and morally correct, I will do for my customer. We need to realize that the marketplace will never be exactly as we predict or want it to be. It is going to evolve and we must be prepared for whatever we get. What I mean by that, specifically, is that in a few years I expect that we'll be reading stories about how the death of open outcry was greatly exaggerated. You'll see a niche on the trading floor, much like what developed at the New York Stock Exchange (NYSE) with respect to its specialist system. Between 1987 and 1989, many specialist firms, thinking the sky was falling when the DOT system was introduced, left the trading floor convinced that the death of the specialist system was imminent. It turned out that they pulled out right before the beginning of the most prosperous period in the history of the NYSE.

QUESTION: *Although your operation is centered in Chicago, Commerzbank is one of Europe's largest banks with a presence in every major financial marketplace throughout the world. Where does the bank see its biggest potential growth markets; in the United States, where the transition to fully electronic trading has yet to take place, or in Europe and the Far East, where electronic trading is firmly established?*

JACK: That's a two-pronged question. There is tremendous opportunity in the United States, especially with a bank FCM that can offer stability and incredible distribution. Nonetheless, our biggest growth market should emerge out of the European theater. Europe, by nature, is underinvested in the equities markets; in fact, only 35 percent of families own equities there, whereas in the United States we have 85 percent participation. Europe is ready to explode with incredible growth over the next 5 to 10 years, especially in Germany. One of the important catalysts will be the soon-to-occur changeover of national currencies to the Euro.

QUESTION: *We both know many floor traders are extremely nervous about the growth of electronic trading. While the floor isn't going away in the foreseeable future, eventually these individuals, if they wish to continue trading, will have to make a transition. How difficult will it be for them to adapt to the change? Which floor trading skills will be most valuable to them in an electronic environment?*

JACK: I suppose it matters what type of trader you are. I guess if you are a pure trader—making decisions based on risk-reward ratios and doing homework—you'll have no problems whatsoever making the transition. On the other hand, if you make your living simply by watching and acting on the order flow in the pit, you will have to try to find new skills to survive. The pit skills will definitely be less valuable over time. In both of those cases, however, the different types of traders add liquidity to the marketplace and we must make sure that during this interim period, when open outcry and technology coexist, that we don't destroy the balance. We need to figure out how to migrate that liquidity to the screen or else everyone is going to suffer.

QUESTION: *You have been a very public proponent of the futures markets, particularly in your frequent appearances on CNBC. In demystifying futures trading you have raised awareness about an endeavor that—while not for everyone—is something that an average investor can include as part of his (or her) trading activity. What are the things that a novice to the futures markets needs to keep in mind in order to have a reasonable chance of attaining success? What role should the broker play in helping the customer become successful?*

JACK: The novice has to understand that futures trading is not a "get rich quick" scheme. It is a money management and risk management skill. Futures trading is something that needs to be learned over time. There are ten thousand lessons to trading. I've been in the business for 18 years and only picked up on a handful of them. I think the smart trader learns something new every day. A trader who feels as if there is nothing new to learn is the trader who usually gets in big trouble. You have to respect the market and never underestimate it. The market is so powerful that no single person is ever larger than it. With respect to the role of the broker, it is his (or her) job to educate the novice. A broker acts as a tutor. You present the customer with the available options, keeping in mind that you have a fiduciary responsibility to him (or her). If a client has a particular strategic objective, it is your job to give him (or her) the direction he (or she) is looking for with as little risk as possible. You want to limit his (or her) risk and increase his (or her) profit potential and show him (or her) how to do it. You also need to make sure that risks are disclosed and that the customer is aware of, and understands, the level of exposure he (or she) is taking upon himself (or herself). I'm actually much less worried about disclosure in an online environment. With an online account, everything is documented and at your fingertips. It's all a question of if you take it or not. But, on the other hand, it is extremely important to help the customer understand all of the aspects of the disclosure form. As the futures markets grow and new participants emerge, it will be a challenge for the broker to make sure that all customers are fully aware of the risks associated with futures trading.

DIRECT ACCESS AND THE NETWORK ENGINEER

Interview with Mike Persico, President, and Pace Beattie, Director of Business Development, Tekom, Inc.

In the "olden days," all you needed to make money on the trading floor was a stack of trading cards, a sharpened pencil, and a good spot in the pit. The infrastructure and technology that supported all of the trading activity remained hidden for the most part, with the exchanges picking up the cost and taking responsibility for implementation. The trader's only worry was figuring out how to buy low and sell high. Today, however, in the electronic age, becoming a successful trader requires a greater degree of involvement in the creation, maintenance, and upgrading of the trading technology. To be sure, the exchanges still are involved and in many ways subsidize the trading activity of both members and non-members of the exchange. Nevertheless, it has become increasingly important to take personal responsibility for ensuring that the hardware, software, and connectivity you utilize allows you to compete on a level playing field. The decisions you make about whether to buy a more powerful PC, which trading front-end to use, or whether to trade with a direct connection to the exchange or over the Internet will dictate, to a large degree, if you make money. To put it in terms that anyone can easily understand, if you were in the NBA you'd wear state-of-the-art sneakers and run rings around anyone who dared to show up on the court barefoot.

Because the vast majority of us are dummies when it comes to technology, we find ourselves intimidated by the idea of entering the electronic future. For most of us, implementing solutions that work and are cost-effective seems beyond our capabilities. It's a new world with a language, citizenry, and culture all its own; one in which many of us feel unwelcome. We know that we need to get ready, but haven't the slightest idea of where to start or whom to trust. We wonder how much it will cost to become competitive. We are vaguely aware that technology is one area in which writing a blank check not only is not a solution, but oftentimes exacerbates the problem. The only thing of which we are certain is that a bleak professional future awaits those who are unwilling to take responsibility for their technology needs.

Mike Persico and Pace Beattie are part of this new world. They are principals in a networking firm called Tekom, Inc. (*www.tekominc.com*), which specializes in trading room build-outs. In fact, I hired them to build and support the two networks Aspire Trading provides in our arcade. The following interview with them addresses many of the key issues you will face as you enter the world of electronic futures trading. One thing I have learned from Mike and Pace in our discussions and business dealings is that while technology can be seen as a potential barrier to success, smart traders will see in those barriers an opportunity to capture a competitive advantage. Because so many traders are likely to make bad technology decisions and because trading is a zero-sum game, those that choose their technology wisely will have the best chance to succeed.

QUESTION: *Mike and Pace, you both have a history of working in the financial services industry, particularly in the derivatives side of the business. Can you tell us a bit about your respective backgrounds?*

MIKE: My name is Mike Persico and I'm the president and CEO of Tekom, Inc. What we are is a full service network, engineering, and systems integration company specifically for the financial services industry. My background is on the IT side of the business. I've been a technician for the last 10 years, and most recently I came from a law firm with 2,000 lawyers and offices all over the world. I was responsible for building and maintaining the networks linking those lawyers and offices together. The networks I built were fast, scalable, and redundant. Now with Tekom, I have customized these types of solutions for the financial services industry. Tekom is a full service provider able to build out any and all types of electronic trading networks.

PACE: My name is Pace Beattie and I'm the business development director here at Tekom. My background is on the trading side. When I came to Chicago in the 1980s I initially worked as a runner and then a phone clerk on the trading floor. I then became a financial futures analyst for a large wire house. I did their broadcast, giving commentary on the markets to their customers, and then I became a floor manager, managing the floor operations for a different firm in the area of financial derivatives. From there I moved into the 30-year-bond futures pit where I traded as a local for 17 years. By mid-2000, I knew that the trading world was going to move from the trading floor to a screen-based environment eventually and decided a career change was in order. So I took my experience and knowledge about trading and looked for a premier technology group that could use my skills. I found Tekom, which was in the process of moving into consulting work helping people create electronic trading environments, and it was a match. I now work for Tekom and make connections with end users who need the kind of services we offer. I go out and talk with traders and companies about what we do and how we can help them accomplish their technology needs.

QUESTION: *Your company helps trading firms deal with their technology needs as they make the transition from the trading floor to the trading room. What are the most common concerns you find that traders are experiencing?*

PACE: Their first fear is: How are they going to make a living? They've traded on the floor for years and their careers and incomes are from the trading floor. They're afraid because the environment they're comfortable with won't be there anymore and they feel like they don't know what to do. That's the biggest concern, and after that it's: Where do I start with electronic trading? How do I do it? How do I get my questions answered? What questions am I supposed to ask? That's where Tekom comes in. We work with these traders and answer all their questions related to setting up a trading room, to exchange issues, to the IT piece.

MIKE: From my perspective, you have people who have done business in a certain way for more than a hundred years and in many cases you have families that have traded on the floor for generations; you'll hear, "My father was a trader, my grandfather was a trader." It's very similar to being a tradesman. When business has been done a certain way for so long, it's only natural for people to fear change. I think a good example of this is some of the agricultural traders at the CBOT. They're probably not the best candidates to move to electronic trading because they've traded on the floor for so long and are not interested in changing. You need to be proficient with the computer—have a comfort level with the mouse and keyboard and with the graphical interface that's on the computer. These guys are going away from an environment they're comfortable with to an environment that they have a distinct discomfort with, so it's a challenging process. I think it's only natural that change of this magnitude generates a lot of fear within the floor trading community. The exchanges are taking measures to try to satisfy both sides of the population with the side-by-side markets we have now. By allowing both systems to coexist—albeit uneasily and at great cost to the exchanges—both those that want to make their living off of the screen and those who remain loyal to the trading floor are able to function. While I think these markets are effective for the time being because of the arbitrage possibilities, the rest of the world is electronic at this point, and Chicago clearly trails behind . . . probably by about three years. While I think eventually we'll be all electronic, this quandary that the exchanges and traders are facing is leading to the slow development of electronic initiatives here in the United States. It's not too hard to understand: trading off of the screen is inherently different than trading on the floor. On the floor your success is based on physical capabilities and proximity to the order flow—the development of personal relationships. Suddenly, on the screen all of that is taken away; you don't have relationships, it's anonymous, and trading becomes more technical. The set of skills that enabled the floor trader to trade successfully in the pit do not necessarily translate well to the screen, so I think that's greatly responsible for the fear these guys are feeling.

QUESTION: *One of the most important decisions a trader will have to make is how to connect to the markets. Traders who need speed and reliability tend to choose direct connections to the exchanges. Traders who are less active may be well-served by an Internet connection. Can you touch on some of the key differences between the two approaches and why professional traders feel a direct connection gives them an edge?*

PACE: I know the most important thing about trading on the screen is that you have to be as fast or faster, if possible, than everyone out there or else you'll get beaten to your trades. And when I go out and talk to traders it's clear that they recognize this. Obviously you need reliability and scalability, but there is no question in my mind that speed comes before everything. The best way to address the speed issue is to have direct connections to the exchanges. This means running lines from the exchanges to your trading shop and having them networked to your screens.

MIKE: I feel that there are three different categories of connectivity to a screen-based trader. I classify them as: 1) direct connection, 2) single pipe service, and 3) the Internet. As Pace stated, a direct connection is a line run directly into an exchange and what that does is put you on a level playing field with all of the exchange members trading electronically. The exchanges have come up with a topology and network architecture that allocates a piece of bandwidth that's the same across the board and that gives you a level playing field so that you are on par with everyone else who has direct access. If you are one off or take a step away from that direct connection, the law of physics dictates that you have another "hop" or segment to travel which will reduce your speed to market. There are 2 things to specifically watch out for here: 1) you do not want your market data prices to be slow, and 2) you do not want your orders and executions to be slow. On the floor there are different ways to get an edge, but on the screen the only way to get an edge is to have speed. If you are forced to trade at less than the speed other traders are operating with, you will lose an edge—the only edge that really counts. In my experience building many networks, direct connection is the best way to get that edge of speed (see Exhibit 2-1).

The next method of connectivity is called *single pipe service* (see Exhibit 2-2). There are firms out there that will have all of the exchange connections in their office and then they will run a circuit into your office and let you choose what parts of their trading network and which exchanges you want to have access to. This allows you to avoid having to spend a lot of money to build the infrastructure to connect with the exchanges and the ongoing costs of maintaining, supporting, and

Exhibit 2-1 Direct Connection to the Exchange

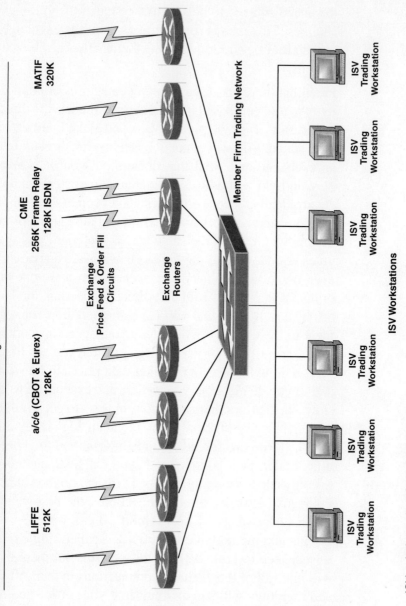

ISV = Independent Software Vendor
Source: Property of Tekom, Inc.

Exhibit 2-2 Exchange Connectivity via Single Pipe Service

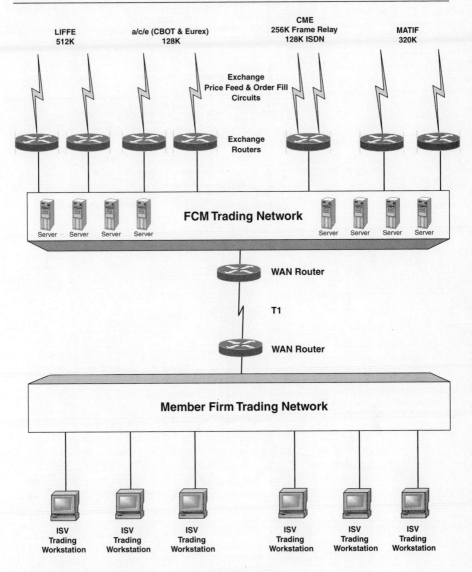

ISV = Independent Sofware Vendor
FCM = Futures Commission Merchant
WAN = Wide Area Network
Source: Property of Tekom, Inc.

Exhibit 2-3 Exchange Connectivity via the Internet

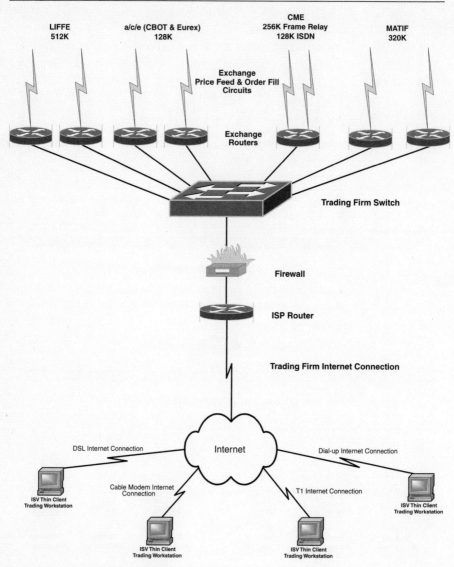

ISP = Internet Service Provider
ISV = Independent Software Vendor

Source: Property of Tekom, Inc.

upgrading the equipment and lines. But it also means that you are one step further away from getting your prices and orders filled expeditiously. There is a balance between cost and speed and ease and speed. It is up to the trader to decide what level of connectivity will help him (or her) best meet his needs and expectations.

The last piece is the Internet (see Exhibit 2-3). We have seen this to be the highest growth area in electronic trading. Historically, high speed capacity on the Internet has not been widely available or cost efficient enough to offer to active futures traders. But now with the advent of broadband connections—DSL and cable modem—high-speed Internet access is becoming available to people at remote locations for a reasonable monthly fee. This allows, for instance, four traders in Naperville (a suburb of Chicago) that may want to reach the market from their arcade to do so. Similarly, a trader at home who may want to simply trade the market after hours can do so with broadband and not be at too great a disadvantage. The Internet has become widespread and dependable enough to allow us to expand this as a medium for trading connectivity. But there are some caveats. The Internet is a public medium and experiences significant general slowdowns if large numbers of people log on. Because of this, Tekom does not recommend Internet trading for professionals. If you want to remain competitive the best way is to be connected directly to the exchanges or connected to a single pipe service from your home by a T-1 line, frame relay circuit, or some form of dedicated point-to-point connection. Although it's acceptable for low volume traders—definitely not scalpers—it's clear that the Internet needs further development before professional traders can use it with confidence. Still, the Internet is only going to get better. One interesting development we are likely to see is the growth of wireless futures trading; soon, you may be able to trade from your boat in Lake Michigan through a hand-held device. I do see that in the not too distant future.

QUESTION: *The CME allows customers to link into its GLOBEX® system through a Virtual Private Network or VPN. How does the VPN connection differ from a standard Internet connection through a broker's order routing system?*

MIKE: There are currently two methods of connectivity for Internet traders. There are Independent Software Vendors (also known as ISVs), who have developed Internet-based applications known as "thin clients." With a thin client, you lose some of the functionality off of the full front-end system that would be on a desktop PC, but you have enough to trade and are able to see a limited depth of market. A VPN, on the other hand, adds an encrypted tunnel such that your connection to the Internet is protected by specific encryption layers which allow the data not to be viewed by anyone with malicious intent. If you don't have this, someone could hack into your Internet connection. Additionally, it allows you to extend your network using the Internet as a medium, which means you can use the full-functionality, or "fat-client" software from your ISV of choice. So the VPN protects your data and it's a much more secure way of trading over the Internet. It does, however, add latency which forces you to sacrifice speed. Because the message must be encrypted on one side and then decrypted on the back side, you sacrifice speed for security. The individual or trading firm needs to decide if the trade-off is worth it.

QUESTION: *So is it fair to say that for an active scalper a VPN solution is probably not appropriate?*

MIKE: We at Tekom have done a lot of testing with particular front-ends and the latency that a VPN client has. I've also sat in front of a VPN client and a direct exchange connection on a PC. The speed differences are down to milliseconds, but milliseconds can make a difference in a trade. From the exchange perspective I understand why they need to protect their networks and cannot allow clear text over the Internet, but, once again, the individual or trading firm must consider the trade-off: speed versus security. This is a decision that's going to be made time and time again.

QUESTION: *If a trader chooses to use the Internet to connect, what should they look for in the Internet Service Provider (ISP)? Can they use services such as AOL, CompuServe, or Prodigy, which are not true ISPs? Also, does it make a difference if they connect to the Internet by way of a dial-up connection, cable modem, DSL Line, or satellite? Are any of these methodologies preferable to the others?*

MIKE: What I commonly recommend for someone who is going to trade over the Internet is a Tier I provider. Tier I providers have very large, high-speed, "fat pipe" connections to the Internet backbone and multiple paths to the Internet backbone for redundancy. Examples of this would be *Ameritech* or *UUNET*—national, almost global, providers of Internet service. If you choose an Internet Service Provider who is regional or a start-up, they certainly aren't going to have the kind of developed infrastructure of a Tier I provider. You need a Tier I provider because they are going to offer better thru-put and higher levels of redundancy. Therefore, if one backbone pipe goes down on the Internet you will have access to another backbone immediately, and your trading will not be interrupted. What I have been told from the front-end software vendors is that the recommended bandwidth for the Internet trader is 128K per individual. That means that a common dial-up connection which typically caps out at around 56K is not acceptable or reliable enough to trade over. DSL is probably the preferred method of connectivity because it is your own dedicated Internet circuit which you do not have to share with anyone else. I do prefer DSL over cable modems because cable modems utilize "shared Internet bandwidth" such that if you are the first person in your apartment building to get a cable modem, then you get the entire amount of bandwidth. But as others come on and join onto the same piece of coaxial cable, your allocation gets smaller and smaller. While a cable modem is better than a dial-up connection, it's still not as good as a DSL connection. One other thing to understand about DSL is that there are two types of connections. One is called ADSL, the other, SDSL. With ADSL, short for "asynchronous DSL," the download feed is higher than the upload feed. Accordingly, you'll get prices in faster than you'll be able to send out orders. I typically do not recommend that type of connection for traders. I like to see them get SDSL, or "synchronous DSL," which gives you an equitable division between your upload and download speeds.

QUESTION: *Independent Software Vendors (ISVs) that provide the front-end trading systems (what the end user sees on the screen and uses to make trades) are able to link their software to the major exchanges using an Applied Programming Interface, also known as an API. What exactly is an API?*

PACE: It's kind of like a handshake between the exchange and the person connecting with the exchange. When the API connection is made it's like the two hands coming together. It's an open code that the software companies can write to so that the ISV's code agrees with the exchange's code and they can communicate back and forth between the trading screen and the exchange.

MIKE: Pace is right. It's like a handshake. It's basically a way for the ISVs to integrate their software into the exchange-mandated architecture. It allows the two codes to interact, for the orders to get to the electronic exchange and the prices to get to the front-end that the trader sees, so basically it's an integration of the interface. It's a spot where the exchange architecture and the third party program touch. Every time an exchange brings its architecture to the markets it needs to have an API.

QUESTION: *As the exchanges increase the distribution of their products through their electronic trading platforms, they have come to rely on ISVs to create front-end trading software. What are the elements of a robust front-end trading system and what should the end-user expect from the ISV in terms of support and maintenance of the product?*

MIKE: While we don't necessarily endorse one ISV over the other—there are a number of very good ISVs and great front-ends—there are certain features that the best of them offer in terms of functionality and support. With respect to functionality, elements like one-click trading, the ability to cancel all orders from the book with one click, and connectivity to many different exchanges have come to be expected by the trading community. As far as maintenance and customer support, I highly recommend you look for an ISV that provides a twenty-four hour support desk, with technicians on the desk who understand both systems and software. This is useful for general informational purposes, and critical in situations where your system goes down and you need to bring it back as soon as possible.

QUESTION: *With the cost of PCs relatively low and seemingly always coming down, should the serious trader buy the most powerful computer on the market in order to achieve a competitive advantage? For example, if two traders have exactly the same skill level and same connectivity, will the trader with a more robust chip powering the CPU or an extra 128K in memory have a better chance of getting the trade?*

MIKE: Electronic trading revolves around obtaining a competitive edge. There are basically two ways to accomplish this: (1) through direct connectivity, or (2) purchasing high-end equipment. If you have a 700 MHZ chip and another trader is working with a gigahertz, he (or she) is likely to have a competitive advantage against you because his (or her) machine will process information more quickly. While it is not possible to quantify exactly how much of a disadvantage the lower-end machine is at, for the small difference in cost between purchasing high-end equipment and whatever Best Buy is selling on the clearance shelf, it makes sense to spend the extra money and put yourself on the most level part of the playing field. The last thing you want to do is put yourself a couple of pegs below your competitors just to save a few dollars. Here's another reason to buy as high-end a machine as you can possibly afford: The shelf life of a PC being used for trading is nine months. To the extent that you can stay ahead of that nine month curve, it will allow you to avoid having to upgrade as frequently as someone buying last year's model. At the very least, make sure that you purchase machines that are expandable. By upgrading you can save some cost by not having to purchase an entirely new computer every time you need, for example, some additional memory. I should also address the same question with regard to networks. I've seen companies spend $150,000 on cherry trading desks and only $5,000 on their networks. It's crazy, because the core of your business is the network you trade on. Be prepared to spend some money to build an ample network and support it. It will be much more valuable to you than a nice looking piece of furniture.

DIRECT ACCESS AND THE TRADER

Paul Doppelt, head of the Doppelt Trading Group, is something of an anomaly. He, and his traders, are successful floor traders that are voluntarily making the transition from the trading floor to the office. The reason they are making this switch, notwithstanding the fact that the floor trading operation is very profitable, is because they see the future and know that it does not include a trading floor.

Interview with Paul Doppelt, Principal, Doppelt Trading Group

QUESTION: Tell us a bit about your background, Paul. How did you come to be a trader?

PAUL: I came into this business in 1987, in July, right before the stock market crash. I went to work for Singer-Wenger. My brother-in-law, Norm Singer, brought me into the business and told me if I was patient he could teach me how to make a living trading commodity options. I worked for them for a few years and then moved off on my own and started trading options in the pit, mostly currencies. After a while I decided to start backing traders of my own, which I have been doing since 1992. At this point there are 14 in our trading group.

QUESTION: *What are the characteristics of a good trader? What are you looking for when you bring a new trader into your group?*

PAUL: That's a very tough question because I've discovered over time that there is no formula that I've been able to discern. There are a lot of things you can look for in a person, such as discipline, curiosity . . . you can look for someone who is studious. But I have the whole gamut of people trading for me now and some have those qualities and some don't. So I'm constantly awed that there is no set formula.

QUESTION: *Do you think anyone can learn to trade if they apply themselves, or is trading an innate skill?*

PAUL: I think anyone can do it, especially electronically.

QUESTION: *Your trading group is very active in the e-mini markets at the CME. One of the elements that distinguishes the e-minis from other electronic contracts is the side-side environment in which it trades with a couple hundred traders trading at workstations that surround the full-sized, pit-traded contracts. Can you tell us about how the traders on the floor—in the pit and on the e-mini workstations surrounding the pit—interact to create this unique symbiotic relationship?*

PAUL: That's a very complex question, even though it appears simple on the surface. The guys trading on the machines initially played off of the activity of the bigger players in the pit. There wasn't a lot of business in electronic futures at the time. People were limited as to the number of contracts they could trade at a time—there was a maximum of 30 that you could enter into the system at once. So the big institutions weren't really participating and volume was low. That gave us an edge. We were able to see trends in the market and we were able to capitalize on that. In fact, some people—and this has grown more prevalent—have put partners on headsets in the pit and do an arbitrage between the pit-traded contract and the e-mini. That approach has grown less lucrative as institutions have entered the market driving the price movement. The pit is no longer relevant. Watching the pit is like looking at the sky and seeing stars that burned out thousands of years ago, but you're still seeing reflections of that light. The pit is no longer the indicator. If you're seeing it in the pit it's too late to act on it. It's already been traded. It's like when you're standing in the pit and an order filler sells you 100 contracts—you've got the edge . . . maybe. The guy in the center of the pit who wants that trade, when he gets it from you a few moments later it's a bad trade . . . for sure. There is a whole world, an electronic world, outside the pit that really drives the market. There are people using our products to trade options at the CBOE, people using our market to front-run their customer orders, people who are using our products against basket trades, micro-sector traders looking at very specific focus groups of stocks who are hedging with e-minis. There are a plethora of users from a hundred different sectors and as a result the edge is no longer there. Day by day, as the electronic trading grows stronger, the pit gets weaker.

QUESTION: *Some perceive that being on the trading floor is a tremendous advantage when trading the e-minis. Yet you have traders both on and off the floor. In using both approaches, are you hedging your bets or preparing for an inevitable transition to a completely electronic environment?*

PAUL: The only reason why my traders are still on the floor is because they have workstations provided by the CME; they have technology that I don't have to pay for. The sooner I can get my guys off of the floor, the better. My off-the-floor traders are more profitable. You have access to much better information off of the floor. In addition to technology, the other reason one stays on the floor is to get access to information. But as the information becomes increasingly less valuable, we have to go off of the floor. Like I said, the people who are driving the market have absolutely no interest in what's happening on the floor.

QUESTION: *The traders in your group are among the most active and successful at the exchange. Without asking you to divulge any proprietary secrets, what types of strategies are they using? Are they technically or fundamentally oriented? How much of their success is due to developing a gut feel for what the market is about to do?*

PAUL: Yes. A bit of everything. David Bixby, my partner and mentor in this market, says that many times the trade you don't make is the best trade. We've come to the realization that it's not how much you trade, but when you trade. It's also knowing when to stay in your trade and how to stay in your trade. Being able to flip out of a trade when you're wrong, without emotion, is a key component of successful trading. People who don't make money get into a series of lousy trades; they hang on too long, and the next thing you know you're out a lot of money. Almost as important is knowing to "load the boat" when you're right. You need to get the trade on and lever yourself into a good position and stay in the trade. Don't get shaken out by the noise. Don't get shaken out by a bit of retracement. It's so important to know how to stay in the winning trades. What most floor traders don't know, and I'll share with you now, is that you cannot live in a vacuum. Our futures products do not trade with their own identity. They are fed by the individual stocks that create them. Those stocks are affected by other stocks and indexes that we don't necessarily trade as futures. Success as an electronic trader will come to those who look at all the information coming from without. You need to digest all the external stimuli and understand it. Information is flowing at a rapid clip. Sitting in a room trading, you want to know every news headline pertaining to the stocks you are trading. Gone are the days when you can simply watch the pit, join the bid or offer, and arb the market. When I first started, I was told you could be trading widgets. "You don't need to be educated," I was told. Those days are gone. You need to understand what's going on at all times. Knowledge counts. Information counts. We meet with our traders every morning before they open. We prepare a summary of all the news that matters, all the charts from the previous day. We analyze the information so that everyone in the group is ready for whatever comes next.

QUESTION: *Most traders do not utilize direct connections to the exchanges, because there are significant up-front and ongoing costs associated with implementing such connectivity. In not being directly connected, one thing they sacrifice is speed. How important is speed? Is a trader who is not connected directly to the markets—let's say someone trading on a broker's order routing system over the Internet—at a disadvantage to the professional traders? Should the average trader be willing to assume additional cost in order to achieve a level playing field?*

PAUL: Absolutely. Absolutely. You may not need or care about the 2, 3, 5, or 10 ticks you're going to get because you're a long-term trader, but the reality is that entry into the market can make your long-term directional play all that more successful. If you have a 5-handle cushion because the speed allowed you to get in at the right time, that's going to give you more leverage, more breathing room for your exit point. Speed is really, really important.

QUESTION: *As the markets and traders become more sophisticated, the lines between the futures and securities world has begun to blur. How important is it that traders learn to understand the relationships between the futures markets they trade and the underlying cash markets?*

PAUL: It's imperative. We carefully watch the cash market when we trade the futures. We know that the cash and futures markets disconnect. The cash market doesn't always represent where the market should be trading. We study the underlying stocks, and so when the disconnect happens we see it and we know why it's happening. The only way to know those things is to be a student of those underlying stocks.

QUESTION: *The day-trading industry has come under criticism because many people who traded successfully during the bull-run of the late 1990s lost everything they made and then some when the market turned downward. Yet in the futures industry, successful traders often make more money in bear markets than they do in a straight-up move. How do you account for this disparity?*

PAUL: We don't have a problem with selling short. You can get short in the futures anytime you want. It's far less complicated than in equities. You can reach down into the book and clear it out if you want. There's no bias in the futures market against being short.

QUESTION: *Many people who try day trading—perhaps even the majority—are ultimately unsuccessful. Why do you think so many fail?*

PAUL: They are not disciplined and they don't really understand what they are doing in the first place. They don't educate themselves. It's really basic: education and discipline.

QUESTION: *We both know that losing is a part of trading and that learning to accept losses and move on to the next trade is one characteristic that distinguishes a successful trader from one who is not. How do you deal with losses?*

PAUL: We don't trade money and we don't trade our positions. We trade the market. On days that we do our jobs, we make a lot of money. On days that we don't, we lose money. How do we look at the losing days? Afterwards, when the market is still fresh in our minds we try to re-evaluate the day. We have a bull session at day's end; we all try to come. We look at charts and constantly surf through and analyze what went wrong so that we don't repeat the same mistakes. Surprisingly, the more we learn about what we do, the less what we think we know matters. The game is changing faster and faster. It used to be that you could figure out a style of trading and get at least a year or two or five where that style of trade would work. That's no longer the case. Things are changing almost daily with the advent of new technology. So we don't think about the money when we're trading. We just try to keep on top of what we're doing and up to speed with the new technologies so that we can continue to move forward.

QUESTION: *Many of our friends and colleagues are scared to death over having to make a transition from the pit to the PC. You are a pit trader who was able to make a successful transition. What advice can you give to those who have trading floor experience, but are reluctant to make the switch?*

PAUL: Change is very difficult. I stood in the Canadian Dollar options pit; a very small backwater pit, really illiquid. I got comfortable there, but the business disappeared. I knew I had to move on, but it was very painful to wrench myself from a place where I had been successful. But it was so refreshing once I made the change and turned on the computer . . . it was almost like being reborn. I found out I am a good trader. I had doubted myself, but once I saw I could do it I moved all of my traders, within six months, from their respective pits onto the machines on the trading floor. Now we're making the transition from the machines on the floor to our offices where we have better information and better technology. My advice to anyone who has been a successful trader and wants to make the transition to trading on a PC is that you mustn't doubt yourself. If you've been successful in the pit you can do it on the screen. You just have to do it.

QUESTION: *Direct access to the securities markets became available in January of 1997 and since that time volume has exploded. When the futures markets are completely electronic and widely distributed, do you think that they will see similar exponential growth?*

PAUL: There's no question about it.

QUESTION: *What are the prospects for Single Stock Futures? Are they the next blockbuster product, or are there too many regulatory hurdles to overcome? The exchanges that are vying for market share in SSFs have suggested that they will create a Designated Market Maker Program in order to promote liquidity. In what ways can Designated Market Makers help these contracts to reach a critical mass of daily volume and open interest?*

PAUL: From what I know about SSFs, they have the potential to be hugely successful. They can change equity trading as we know it today and they are just the beginning. The only way, however, SSF will flourish is with a Designated Market Maker program. The liquidity has to come from somewhere and it is the market making community that has the expertise and capital to provide it.

QUESTION: *Any final thoughts about electronic futures trading?*

PAUL: I want to say that investing in technology infrastructure is the key; not just to be able to trade one product, but to be able to trade a whole plate of products. It's extremely expensive to build and maintain technology. It's been a great lesson for us, dealing with the costs and headaches; learning just how little we actually know about it. I would suggest that anyone who is thinking about electronic trading should align themselves with an organization or trading room that will have state-of-the-art technology. If you want to succeed at this, you need to be no slower than the fastest person getting into the market and no slower than the person who receives information fastest. We're not just competing against people anymore; we're competing against programs. It's like when you stand in a pit, you want to be next to the broker; here you want to be close to the information. In order to do that, you need the best technology.

THE TRADE MATCHING ENGINE: GLOBEX2—A CASE STUDY

The CME's GLOBEX2 (G2) is an example of an exchange-created and supported trade matching engine. Independent Software Vendors write the sexy programs that traders see on their monitors, but it is the exchanges that provide the foundation of the market with their trade matching engines. A trade matching engine is *the place where all trades take place and from which market data is disseminated.* The Chicago Board of Trade and EUREX trade on the *a/c/e* platform. The New York Mercantile Exchange (NYMEX) uses a system called *ACCESS;* the Sydney Futures Exchange (SFE) trades on a system called *SYCOM;* and the London International Financial Futures Exchange (LIFFE) trades on its system, known as *LIFFE CONNECT.*

Each of the exchanges argues that its system is superior: faster, more reliable, and easier for end users to master. Of all the exchange systems available, I am most familiar with G2, having worked on one of the committees that developed GLOBEX in the early 1990s, through my personal trading activity, and through the trading activity of my customers. While I cannot settle the argument here as to which system is the best, I have been consistently impressed with the G2 matching engine. As volume has increased dramatically (as of April 2001, CME electronic volume was up 143 percent from the prior year), the system has kept up remarkably well. It is available 23 hours per day, Sunday night through Friday afternoon (Chicago time). It is extremely reliable and, on the rare occasions when there has been a systemic problem, the exchange has done a great job of both informing the user community and solving the crisis. G2 has always been speedy, but recently the exchange broke the "one-second barrier," which is like running a sub-3:50 mile for the first time. What this means is that the current round-trip response time, which includes order entry, matching, and execution report averages between 0.5 and 0.7 seconds.

In order for you to better understand what a trading engine is and how it functions, consider the following case study of the components of G2. In it, you will find the system dissected into various pieces that make up GLOBEX and comprise its functionality. While no two matching engines are alike, this study will give you a good idea of the features contained in a premier system.

Order Types

The trade matching engine accommodates the following order entry types.

Bid/Offer

An order type where the buyer defines the maximum price to pay (Bid) and the seller defines the minimal price to sell (Offer).

Limit Order

The Limit Order to buy (sell) remains in the book until the order is executed or cancelled. When this order is entered, it is executed (if possible) at the limits of the orders already present in the order book.

Bid-Ask

A Bid-Ask order allows for simultaneous entry of a two-sided market (Bid and Offer/Limit orders).

Market Order

A Market Order is an order to buy or sell at the best possible price in the market at the time of order entry. Any remaining quantity will become a limit at the execution price.

Stop Limit

A Stop Limit order is executed within a given price interval as it is entered with a trigger price and a limit price. For the buy (sell) order, the trigger price is lower (higher) than the limit price. To be valid, a buy (sell) Stop Limit order must have a trigger price greater (lower) than the last traded price for the instrument.

   The order can be executed within the limit price once the order is placed in the order book after the order's trigger price is traded on the market. The order will be executed at all price levels between the trigger price and the limit price to attempt to fill the quantity specified in the order. If the order is not fully executed, then the remaining quantity of the order is left in the system at the specified limit price.

Stop Loss

A Stop Loss order is executed at the best available opposite side limit price after its trigger price is elected. To be valid, a buy (sell) Stop Limit order must have a trigger price greater (lower) than the last traded price for the instrument.

Market On Opening (MOO)

A MOO order is entered during the pre-opening staging period and is only executable at the Opening price.

Market On Close (MOC)

A MOC order is entered any time during the ETS trading day and is only executable at the Closing price.

Order Cancels Order (OCO)

An OCO order is a combined order. The two legs are entered at the same time in the system, but the two legs are not traded simultaneously. Instead, if one order is traded in whole or in part, the other is cancelled entirely. The two legs for OCO orders are Limit and Stop/Limit orders.

All Or None (AON)

An AON order must execute all at once or not at all. If there is insufficient quantity on the opposing side of the market to match, the order is placed in the Order Book.

Display Quantity

When the Display Quantity function is used, only the quantity entered in the display field is shown on the Order Book. Once it has been filled (possibly in several trades), a new displayed quantity is shown at the book's lowest priority. The display quantity refresh continues until the total (hidden quantity) has been filled.

Minimum Quantity

At the time of order entry, if the minimum quantity indicated does not match, the whole order is cancelled. If the minimum quantity does match, the remaining quantity stays in the order book and can match with any new incoming orders. If the Minimum Quantity equals the total orders quantity on the opposing side of the market, then the order will act like a Fill or Kill order.

Order Qualifiers

The host will accommodate the following order qualifier types.

Session/Day

An order with a Session qualifier is only valid until the end of the current intraday session. At the end of that period, the order or its remaining quantity is cancelled and purged from the central Order Book.

Fill and Kill

This order is filled upon introduction to the system, if possible. If the order is not executed totally, the system cancels it. If the order is only filled partially, the remaining quantity is cancelled.

Fill or Kill

This order must be fully executed after its introduction in the system. If it is not executed for the total quantity, the whole order is cancelled.

Good 'Til Canceled

This order will stay valid until it is cancelled by the trader or until the underlying contract expires.

Good Through Date

This order will stay valid until it is cancelled by the trader or until the end of the day specified by the trader upon order entry.

Clearing Information

U.S. Regulatory Clearing Data

The following CFTC mandated information is included within every order/trade message: Account Number, Origin Type, and Customer Type. Additional clearing fields are supported that are not CFTC required fields, including: Fee, Memo, Give Up, and Posting Action. Order messaging relating to the creation, modification, and acknowledgment of orders also includes the clearing data. In addition, every trade acknowledgment for each trade counterpart includes the clearing data of the given counterpart plus the firm ID/trader ID of the counterpart.

Give Ups

Upon order entry, an executing firm/member may enter an order on behalf of a different firm/member. The order is specified with a Give Up and clearing firm number in the clearing data section of the order.

Host Order Number (HON)

The HON is automatically attributed by the host for every order per firm, per instrument, and per trading day.

Host Trade Number (HTN)

The HTN is provided by the host for every trade per instrument and per date. The HTNs can be reinitialized to zero at the end of a trading day or week.

Book Display

The matching engine accommodates the following book display types.

Price Book

Displays a 5-deep market (5 best-bid and 5 best-offer prices) with order quantities aggregated at each price.

Market Book

Displays all individual orders for the entire book.

Order Management

The matching engine accommodates the following order management features.

Order Cancellation

The host can accommodate order cancellation by any of the following ISV-driven methods:

- Single Order
- All Orders
- All Bids
- All Offers
- By Instrument
- By Group
- Contract
- Account

Orders cancelled by market surveillance will be flagged as such by the host.

Order Modification

The host allows for the ability to modify any field of a resting order. When modified, an order will lose its priority upon the following cases:

- Increase in Quantity
- Change in Price
- Account Number Modification

In all other cases, the priority will remain the same.

Access to the Trade Matching Engine

Applications Programming Interface (API)

The host allows for interfaces from a firm's proprietary or third party system via an applications programming interface (API) using the Financial Information Exchange (FIX) protocol. These protocols establish a connection with the host using preset messaging which allows third party software to access the trading engine and enter orders.

Spreading

The matching engine will accommodate the following spread types.

Exchange Defined Instrument

Single instruments defined as a spread. The host will accommodate positive, negative, and zero prices.

Buy/Buy (Sell/Sell)

Single-sided multiple-spread legs (up to 24 legs).

Bundles/Packs/Strips

Multi-legged Exchange-defined spreads ranging up to 40 legs.

Option Strategies

Exchange-defined option-spreading strategies including:

- Straddle
- Strangle
- Vertical
- Horizontal
- Diagonal
- Strip
- Risk Reversal
- Butterfly
- Iron Butterfly

Non-Trading Functions

Request For Quote (RFQ)

The host will accommodate any user to send a RFQ for Commodity and Quantity (optional). Such messages are broadcast by the host so that the trading population can answer by issuing a relevant buy and/or sell order.

Ticker Display

G2 displays all ticker messages processed by the host. This includes better bids (higher), or better offers (lower), and all trades executed. The ticker messages will also display the Indicative Opening Price values during the Opening phase of trading.

Audit Trails

The host accommodates the following audit trail and logging capability.

Host Audit Trail (HAT)

The HAT contains all activity for all users/workstations throughout a session.

Host Reports

Various statistical and user activity reports from the Host Audit Trail.

Anonymity of Orders

The host accommodates the following market Transparencies.

Pre-Trade

All orders are completely anonymous although the host can accommodate broadcasting the Firm ID and/or Trader ID. Market Surveillance can always ascertain the order originator regardless of the anonymity method chosen.

Post-Execution

All executions are anonymous to the trading population. However, contra-party information (firm and user) is transmitted on trade reports to parties of the execution. After a match, the host sends three different messages:

1. Broadcast information to the trading community (summary of the trade without clearing information).
2. Trade acknowledgment back to the counterparties (includes all information except the clearing information of the counterparty).
3. Trade message to clearing system (all information on the trade including all clearing data for both counterparties).

Processing Rules

The engine accommodates the following order matching algorithms.

First-In First-Out (FIFO)

The FIFO method uses price and time as the only criteria for filling an order. When trades are matched using the FIFO method, orders at the same price level will be filled according to time priority.

Allocation Algorithm

The Allocation Algorithm matches orders based on price, time (for the first order only that "betters" a market), and size.

The Allocation Algorithm operates according to the following rules:

1. Orders placed during the "pre-opening" or at the indicative opening price (IOP) will be matched on a price and time priority basis.
2. Time priority is assigned to an order that betters the market (i.e., a new buy order at 36 betters a 35 bid). Only one order per side of the market (buy side and sell side) can have time priority. There will be situations where time priority does not exist for one or both sides of the market (for example, an order betters the market, but is then cancelled), but there will never be a situation that results in two orders on the same side of the market having time priority.
3. Time priority orders are matched first regardless of size.
4. After a time priority order is filled, the Allocation Algorithm is applied to the remainder of the resting orders at that price level.
5. The Algorithm allocates fills based upon each resting order's percentage representation of total volume at a given price level. For example, an order that makes up 30 percent of the total volume resting at a price will receive approximately 30 percent of all executions that occur at that price. We say approximately because any fractions (of lots) resulting from the percentage allocation are dropped (rounded down) and not allocated by this rule.
6. After percentage allocation, all remaining lots not previously allocated (in other words, the lots "dropped" from the step above) are allocated to the largest order remaining at the traded price. If two or more orders have identical quantities and are the largest orders, GLOBEX2 will perform an "electronic coin flip" to determine the order that receives these remaining lots.

The four examples that follow describe how orders are filled by the numbered allocation rules defined above. The market is considered open so rule #1 will not apply. All examples are independent of each other.

Order Ref.	Bid Quantity	Bid Price	Ask Price	Ask Quantity	Order Ref.
A	100**	36	37	150**	E
B	250	36	37	100	F

Order Ref.	Bid Quantity	Bid Price	Ask Price	Ask Quantity	Order Ref.
C	250	36	37	300	G
D	500	36	37	300	H

**These orders have time priority.

Situation 1: Order X is placed to buy 1 @ 37. Order X trades with Order E because E has time priority (rule #3) at 37. E now has 1 fill and 149 still working. Time priority is still maintained by E for the balance of the order (149 lots) until a sell order betters the market (rule #2). Furthermore, because X bettered the bid side of the market (rule #2), order A loses time priority and any trades that occur at 36 will be allocated by algorithm for all orders at that price.

Situation 2: Order X is placed to buy 200 @ 37. X trades with Order E, which is completely filled because it had time priority at 37 (rule #3). The remaining 50 lots from X are allocated by algorithm (rule #4 and #5):

- F is filled on 7 lots
- G is filled on 21 lots
- H is filled on 21 lots

Allocation by algorithm is now complete, but because of rounding there is still 1 lot to distribute (rule #5). G and H are both the largest orders, so GLOBEX2 performs an electronic coin flip to determine which order gets the extra lot (rule #6). In this example, G wins and ends up being filled on a total of 22 lots.

Lastly, because X bettered the bid side of the market (rule #2), order A loses time priority and any trades that occur at 36 will be allocated by algorithm for all orders at that price.

Situation 3: Order X is entered to sell 450 @ 36. X trades with Order A, which is completely filled because it had time priority at 36 (rule #3). The remaining 350 from X are allocated by algorithm (rule #4 and #5):

- B is filled on 87 lots
- C is filled on 87 lots
- D is filled on 175 lots

Allocation by algorithm is now complete, but because of rounding there is still 1 lot to distribute (rule #5). D is the largest order, so

GLOBEX2 allocates the extra lot (rule #6) to D. No electronic coin flip is needed because there is only one largest order. D ends up being filled on a total of 176 lots.

Lastly, because X bettered the ask side of the market (rule #2), order E loses time priority and any trades that occur at 37 will be allocated by algorithm for all orders at that price.

Situation 4:   Order X is entered to sell 500 @ 36. X trades with Order A, which is completely filled because it had time priority at 36 (rule #3). The remaining 400 from X are allocated by algorithm (rule #4 and #5).

- B is filled on 100 lots and as a result works 150
- C is filled on 100 lots and as a result works 150
- D is filled on 200 lots and as a result works 300

Allocation by algorithm is now complete, and rounding was not an issue.

Lastly, because X bettered the ask side of the market (rule #2), order E loses time priority and any trades that occur at 37 will be allocated by algorithm for all orders at that price.

GLOBEX2 Trading Phases

Trading Phases are managed according to the instrument group (class). An instrument group is usually part of the same market segment, where instruments share the same basic rules and/or the same timetable template.

The matching engine accommodates the following trading phases.

Beginning of Day

During this initial phase, the Order Book may be viewed, but no order entry, modification, or cancellation is allowed. The Control Center utilizes this phase to ensure all trading schedules and to make adjustments if necessary.

Pre-Opening

This phase allows traders to enter, modify, or cancel orders. However, Hit, Take, and Market orders are not permitted. Orders introduced during this period contribute to the calculation of the IOP, but are not traded. The IOP represents the price at which an instrument will trade or what the opening price will be on the Bid or Offer side when the instrument begins trading. The IOP is updated real time throughout the entire IOP period. The basic rules for calculating the IOP include:

- Maximization of matching quantity
- Minimization of non-matching quantity
- Highest price used if non-matching quantity is on the buy side for all prices
- Lowest price if non-matching quantity is on the sell side for all prices
- Closest price to Reference price

Pre-Opening/Non-Cancel Period

For 30 seconds before the session opens, traders can enter orders, but are not permitted to Cancel, Modify, Activate/Deactivate, Hit/Take, or enter Market orders. Indicative prices are calculated and broadcast as changes occur.

Deferred Opening

The host will accommodate a deferred opening by instrument in the event of a market emergency. As such, predefined trading phases indigenous to the opening can be delayed en masse according to a time parameter that is administered by the Control Center.

Market Opening

The system calculates the opening price. All orders that can be executed at this opening price are traded. Orders entered during this period are queued and then processed when the market switches to the Continuous Trading phase. Normal trading begins at the conclusion of this phase.

Continuous Trading

This phase is triggered as soon as the Market Opening phase ends. The switch to the Trading Session phase (Continuous Trading) marks the end of the Opening processes. Orders are sent to the market real time based on the instrument's trading times.

Call Auction

A Call Auction is a market where orders are grouped together for simultaneous execution, in a single multilateral trade, for a specific product, at a prescribed time, and at a prescribed price. The price is the value that best equates the aggregated buys and sells. Buys at this price and higher and sells at this price and lower generally execute.

If, because of quantity discrepancies, an exact match between aggregate buys and sells does not exist at any price, then buy orders placed at the clearing price do not execute in full (if buys exceed sells) or sell orders do not execute in full (if sells exceed buys).

Time priority or pro-rated executed (equal percentage of each order executed) may be used to determine which orders to execute among those that have been placed at the lowest executable bid (if buys exceed sells) or at the highest executable ask (if sells exceed buys).

Surveillance Intervention

Trading is closed with only the Control Center having the capability to process certain commands.

End of Session

Electronic trading is closed.

Post Session Batch or Mini-Batch (for 24 hour trading)

The following host-directed system events occur:

- modification of trading date
- purge of invalid orders
- resetting of statistics (e.g., high, low, volume)
- updating of reference data and member information
- generation of files for historical data

Trading Timetables

Timetable templates are defined for each instrument group. A template is a succession of programmed phases that make up the trading day. A template can span over two or more calendar days.

Instrument States

The general state of an instrument can be Authorized or Forbidden. The Authorized state allows for order entry, modification, and cancellation. The Forbidden state disallows order entry, modification, and cancellation.

There are also temporary instrument states:

- *Frozen*—Order entry, modification, and cancellation is temporarily denied on the instrument. The GCC can reject the order or accept the order (thaw), causing the market to freeze.
- *Reserve*—Order entry, modification, and cancellation are permitted for the instrument, but order execution is denied (like during Pre-open).
- *Open*—Order entry, modification, and cancellation are allowed for the instrument, assuming the group state permits.
- *Suspend*—Prevents order entry, modification, and cancellation for an instrument.
- *Halt*—Order entry, modification, and cancellation for a group or market is allowed, but order execution is denied.

Price Calculations and Support

The trade matching engine accommodates the following price calculations and support.

Price Limits by Contract

Each Contract has its own price limits (High and Low). Orders placed outside the limit range will be rejected by the host with the relevant error message.

Price Conventions

The host currently supports the following price types: Whole, 1/4 32nds, 1/2 32nds, 32nds, 1/2 64ths, 64ths, 1/2, 1/4, 1/8, 1/16, and Decimal (decimal point displayed).

Different Tick Values by Contract

Each contract has its own tick values.

Different Tick Values by Price Levels

Tick values can be set at various price levels. For example, the tick value decreases as the price of an option gets below a specified level.

Cabinet Prices

Nominal price for liquidating deep out of the money option contracts; defined as the lowest possible tradable price (below the minimum tick value and determined within the clearing System).

Trade Cancellation

The host accommodates the ability to bust error trades with the removal of transactions from the trade database, and automatic resetting of market statistics (Last, High, Low, Volume).

The host will accommodate the following database maintenance functions performed by the Control Center:

Contract Maintenance

The ability to Add, Modify, and Delete contract profiles online or offline (batch mode). All offline trading system modifications (batch or online) to the contract database can be downloaded to the firm servers at system start-up or on server recovery so that all workstations are in sync.

Contract Profile

Contract profiles contain the following elements:

- Member Class
- Contract Name
- Contract Type
- Price Limits (High/Low)
- Settlement Price
- Tick Values
- Ticker Codes
- Strike Price (if necessary)
- Lot Size
- Minimum Quantity
- Underlying Contract
- Settle Date
- Expiration Date
- Activation Date
- Minimum Quantity/Maximum Quantity
- Algorithm

Modification of Contract Settlement Price and Limits Online

Modifications can occur by individual contract (online) or for all contracts.

Referential Data

Modifications can occur by individual contract (online) or for all contracts for settlement price and high/low limits. These modifications can occur either automatically via referential data sent by Clearing or by Control Center manual intervention.

Reset Statistics

At the end of a session, a global statistic reset will occur.

Security

When a SLE connects to the Host Frontal, the corresponding firm ID is checked for validity. In addition, a correct password must be provided by the SLE. The verification of a connecting SLE against its IP address is also performed by the Frontal.

Show Active Users

The host will send alert messages for all user logon and logoff activities. In addition, the host will show all users logged onto the system.

View Orders and Trades

The host accommodates the Control Center to globally and/or selectively view orders and trades.

Trade Cancellation for Error Trades

The host will accommodate the ability to cancel trades with the removal of transactions from the trade database and automatic resetting of market statistics (Last, High, Low, Volume).

Broadcast Messages

The host will accommodate the Control Center to send broadcast messages to all firms and workstations.

Market Move Alerts

Market move alerts on a given contract will be generated in the following conditions: in X percent of the incoming trade versus Last, new Last versus previous Last, or incoming IOP versus Last or settlement price.

Order Entry Alerts

An alert threshold can be generated when a Last or IOP occurs at a configurable percentage below the high limit or above the low limit. In addition, alerts on entered quantity can be generated as well as alerts on volume (price times quantity).

Market Transition

Market transitions can occur automatically as defined by schedule management or manually as prescribed by a Control Center command. Market emergency situations are also accommodated by allowing the

Control Center to interrupt the entire market, a given class, or a given instrument.

Price Banding

The host will reject erroneously entered orders with exceedingly unreasonable prices. "Unreasonable" prices are prices that exceed an absolute value as configured by the Control Center. Contingent conditions include Market State, underlying trading conditions (i.e., prevalence of IOP and/or Last for a given trading state), and price (settlement, Last, or IOP).

Quantity Restriction

The host will reject erroneously entered orders with exceedingly unreasonable quantities. "Unreasonable" quantities are quantities that exceed an absolute value as configured by the Control Center.

Upcoming Enhancements to the System

The following enhancements to the trade matching engine are currently in development and are targeted for deployment by the first quarter of 2002.

185 Transactions Per Second (TPS)

G2's current performance of 185 TPS will be increased to 500 TPS.

Implied Spreading

An Implied Spread is a spread that is created from individual outright orders that are available in the marketplace. Implied IN/OUT spreading is where the trading engine will work a spread order in the outright contracts and work outright orders in the spread market at the same time without the risk to the trader/broker of being double filled or filled on one leg and not on the other leg. Implied spread trading is utilized to provide liquidity to the existing market. For example, a buy in one contract month and an offer in another contract month of the same futures contract can create an implied market in the corresponding calendar spread.

There are two different ways of processing implied spread orders:

1. Implied <<**IN**>> orders are derived from existing outright orders in individual contracts (legs). This means that an outright order on a spread can be matched with other outright orders on the spread OR with a combination of orders on the legs of the spread based on price and time priority rules.
2. Implied <<**OUT**>> orders are derived from a combination of existing orders on the spread and existing orders on the individual legs. One leg of the spread is utilized to create a contingent outright order on the other leg of the spread. This means that an outright order on a leg can be matched with other outright orders on this specific leg OR with a combination of orders from *any* spread composed of this leg and orders of the other legs of the spread based on price and time priority rules.

Priority Allocation Algorithm

This algorithm is specific to a given Trader ID. A flag and percentage indicator would be required for the Trader ID at a group level. This flag and percentage will allow the individual trader to be allocated a larger percentage of the allocation based on the percentage amount that is associated with an incentive percent that is preset for the Trader ID. This percentage will be configurable by Trader ID. The percentage indicator will be applicable to a given range and will be able to be changed online during the trading session. In addition, the same rules regarding Allocation Matching will also apply to the Priority Allocation Algorithm. In situations where a display quantity is utilized on an order, the Priority Algorithm multiplier will only be applied to the display quantity of the order, not the total quantity. The multiplier will continue to be added as the display quantity is refreshed. An example of the Priority Allocation Matching Algorithm is given:

There are a total of four orders with different quantities on the book with the same price. Only one order is from a user with a priority multiplier of more than one (Order No. 2). An inbound order to match has a quantity of 130:

Order Number	Quantity	Multiplier	Weighted Quantity
1	50	1	50
2	40	2	80
3	30	1	30
4	20	1	20

Percentage to allocate (quantity of the order divided by the total quantity of orders):

Order 1 = 50/180 × 130 = 36
Order 2 = 80/180 × 130 = 57 (40 is the maximum allowed)
Order 3 = 30/180 × 130 = 21
Order 4 = 20/180 × 130 = 14

On the first pass, Order 2 would be allocated 57.77 (80/180 × 130) or up to as much as the multiplier of 2 multiplied by the quantity of Order 2. Since it is not possible to overfill this order, the following calculations are utilized:

1st pass:

Order Number	Quantity	Multiplier	Weighted Quantity	Quantity Filled
1	50	1	50	36
2	40	2	80	40
3	30	1	30	21
4	20	1	20	14

111 lots have been filled with 19 remaining to be allocated to the largest order remaining at the traded price in the 2nd pass, still using the weighted quantities instead of the original quantities for distribution:

Order Number	Quantity	Muliplier	Weighted Quantity	Quantity Filled	2nd Pass Filled	Remainder
1	50	1	50	36	14	–
2	40	2	80	40	–	–
3	30	1	30	21	5	4
4	20	1	20	14	--	6

3
THE CASE FOR FUTURES TRADING

EVERYBODY NEEDS INSURANCE

To one degree or another, everyone is terrified of the future. Since no one knows with complete certainty what the next minute will bring, much less where the stock market will be in three months, we worry about how to protect ourselves. We try to anticipate the worst case scenario. We buy insurance: health, disability, life, automobile, home, mortgage, liability, and for the truly anal, policies that provide protection in the event that, say, an eye or leg is lost while flying in a commercial jet. While some insurance is a sucker's bet—nobody really needs to buy term-life insurance for a newborn—we cannot dismiss the importance of trying to hedge our bets; of planning ahead so that at the end of a hard day sleep comes easily. That is why there are futures markets: to allow buyers and producers of goods to pass along the risk that something awful might happen in-between the time that they need to buy or are ready to produce.

Although relatively few will ever trade in these markets, they have a profound effect on everyone in the world. The price for virtually everything that is consumed is inextricably linked to the prices set in the pits and trading rooms in Chicago, London, Frankfurt, Tokyo, and a hundred other trading centers throughout the world. If you shop for food; carry a mortgage; heat your home; fill up your car with gas; buy jewelry; use a credit card; own stocks, mutual funds, and bonds; or even if you just like a glass of frozen orange juice, a bowl of Frosted Flakes, and a couple of strips of bacon with your breakfast, you are affected by the futures markets.

HEDGING 101: THE BASICS

In order to understand the influence that these markets have on our lives, it is first necessary to explain how they work. A futures contract is an agreement between two parties, a buyer and seller, to set a price on a fixed amount of a commodity for a delivery date in the future. On the delivery date, the buyer pays the agreed-upon price and the seller delivers the agreed-upon amount of the commodity. Between the time the parties agree to the transaction and the subsequent payment and delivery, the price of the commodity may fluctuate. So, for example, if the deal calls for the buyer to pay $5 for a box of widgets to be delivered in 90 days and at that time the price of widgets has fallen to $4 a box, one might argue that the buyer made an ill-conceived deal: if he had only waited, he could have saved 20 percent. On the other side of this deal, it would seem that the seller must be one shrewd widget-maker. In a market where prices have plummeted, he will receive a 20 percent premium. What if the price of widgets had risen to $6 per box during the 90 days between the transaction and delivery? Would we then praise the buyer's trading acumen and consider the seller the dunce?

In fact, both parties were smart in both scenarios. By entering into a futures contract, the buyer agreed to pay a price that he could afford for the widgets he needed in 90 days. In contrast, the seller knew that in 90 days he would receive a price for his widgets that justifies the costs he bears in producing them. Neither party particularly cares whether the price of widgets moves up or down once they enter into their transaction, because at a price of $5 per box both ensure they can operate their respective businesses at a profit. If you follow this logic, you now have a basic understanding of why the futures markets exist: *by passing off an unwanted risk to a counterparty who is more willing or able to take that particular risk, a hedge against the uncertainty of the future is created.*

Let us consider an example of how this works in the real world. An automobile dealer in Anytown, U.S.A., will be receiving a shipment of 100 Mercedes convertibles from Daimler-Chrysler in 90 days, each of which has been pre-sold for $100,000. The dealer has built in a profit margin of $20,000 per unit and anxiously waits for the day that the cars arrive, so that she can deliver them to her customers and pocket the gain. The deal though is not quite that simple. Daimler will not accept dollars for this transaction. Therefore, in order to pay for the shipment, the dealer must purchase $10 million worth of eurocurrency (euros) sometime prior to the end of the 90 days.

Just as the price of physical commodities like gold, wheat, or oil fluctuates, the value of one currency relative to another can vary. If, for example, a country has volatile interest rates, is politically fragile, or employs restrictive trade regulations, the value of its currency is unlikely to remain stable. Our auto dealer knows that any of those, or a thousand other unforeseen factors, could change the relative value of the dollar to the euro. We will assume for purposes of this example that the euro is equal to exactly one dollar at the beginning of the 90-day period. If the euro appreciates 10 percent against the dollar by the delivery date, the dealer will have to come up with an additional $1 million to buy the euros she needs to pay for the cars. As a result, she will see her anticipated profit of $20,000 per convertible slashed in half. How can she avoid this potentially devastating outcome and protect her hard-earned profit?

By buying futures, of course! If, with the euro and dollar trading at the same value, she purchases futures contracts for $10 million worth of euros, she will lock in her $20,000 profit per car irrespective of what happens to the European currency prior to payment and delivery. If the euro appreciates, as in the aforementioned example, she will not be affected because the counterparty that agreed to sell her the futures contracts will deliver the euros at the lower price on delivery day. The dealer does not care whether the price of euros depreciates during the 90 days—enabling her to pay less than $10 million for her euros—because she is an auto dealer, not a currency speculator. Her only goal is to be able to pass off the risk for 89 worry-free days and restful nights and know that on the glorious morning of that ninetieth day she will wake up and collect her $20,000 per-car profit.

Our friend the auto dealer is not the only party concerned about what might happen to the value of the euro during the 90 days. In the event that the currency depreciates by 10 percent, Daimler will receive only $9 million worth of euros. As the producer they have the opposite, but equally significant, concern that troubles the buyer. If they don't manage their exposure to an adverse currency move, they may have to say "auf Wiedersehen" to their profits. They can protect themselves, however, by selling euro futures contracts worth $10 million. By doing so, they ensure that 90 days hence they'll receive an amount that satisfies their business needs irrespective of the value of the euro on the day the cars are delivered and the auto dealer pays for them. Daimler does not care whether the euro appreciates: the windfall that they would receive in such an event—had they not hedged their exposure—comes at too great a risk. Like the auto dealer, they are not currency speculators; they are just trying to sell their cars at a profit. These examples are simple hedges and leave out a lot of the technical details that would be considered before determining if a hedge is cost-effective, such as the interest rate expense of carrying a position for 90 days, volatility, default risk, and whether the entity wants to hedge the entire exposure or only a portion of it (even hedgers sometime get the itch to speculate). Nonetheless, this example of a classic hedge is illustrative of the basic concept of, and justification for, the existence of the futures markets.

THE SPECULATORS

This brings us to the role of the speculator. While it might be difficult to convince anyone that traders who fall into this category perform a noble function, it is reasonable to say that if the speculator did not exist it would be virtually impossible to know the fair price for any good or service. The speculator is a mercenary, committing capital to positions to which he has no particular affinity as he is neither a natural buyer nor seller. Rather, by providing a continuous bid-offer spread—known as *market-making*—he bets that he can buy low and sell high from nervous hedgers—the *market-takers*—who will pay almost any price to avoid risk. This constant market-making creates liquidity, which is the grease that lubricates the market's movement. Although the speculator produces nothing—except perhaps trading profits—his role is no less important than that of the hedger. Without liquidity, economic activity grinds to a halt. In fact, in emerging markets where liquidity is scarce it is difficult to conduct business. The absence of liquidity can have dire consequences, as the world discovered during the Orange County, Mexican Peso, and LTCM debacles of recent years.

The speculator is the "anti-hedger," thriving on the volatility that hedgers abhor; finding value in situations from which hedgers shy away. In short, the goal of the speculator is the diametric opposite of the hedger: *The speculator establishes a position in the market and tries to profit from market movement.*

ORGANIZED FUTURES EXCHANGES: AN ABRIDGED HISTORY

Futures markets have existed for millennia. In the Biblical Book of Genesis, Joseph interprets the Pharaoh's dream: seven fat cows devour seven emaciated cows. Joseph, in what must be among the first recorded instances of fundamental analysis (not to mention insider trading), tells the Egyptian king that the palace must collect grain during an upcoming seven years of plenty and store a portion of the bountiful harvest to satiate the needs of the people during a subsequent seven years of famine. When the famine came, Egypt prospered as supplies diminished and prices rose. Pharaoh had speculated and profited spectacularly when the starving peoples of the region were forced to come to Egypt and pay a premium to buy the food they needed. Had there been an Egyptian Mercantile Exchange, perhaps some would have bought and sold futures contracts and been protected, but organized exchanges did not come into being until sometime in the sixteenth century. At that time, in Japan, a market for rice futures was created, to allow merchants the ability to deal with erratic supplies of the commodity resulting from a series of civil wars among the Samurai. Later, in seventeenth century Holland, a futures market was created for trading in tulip bulbs that were all the rage throughout Europe. The "tulip-mania" which fueled the market eventually expired. Many speculators lost fortunes and Europe was plunged into a deep financial crisis. These attempts to create reliable risk management marketplaces were primitive by comparison to what we have today, but underscores that concerns about risk and the need to control its pernicious effects have existed for as long as traders have brought their goods to market.

In 1841, the Chicago Board of Trade was created by a group of grain merchants, and the era of the modern exchange commenced. Innovations like the *clearinghouse*, a central organization that ensures traders fulfill their commitments to each other; *marking to the market*, or accounting for profits and losses at the end of every trading day; and the concept of *margins,* or the collection of good faith deposits to collateralize positions, allowed the exchange to earn the confidence of investors and grow to prominence. The Chicago Mercantile Exchange came into being as a place to trade butter and eggs. For most of its history, the exchange maintained a low profile, although it attracted attention in the 1950s when it was involved in a scandal involving the manipulation of onion futures, and in the 1960s when it successfully introduced futures markets for live cattle and pork bellies. Still, although the exchange was growing, it remained on the back burner until the late 1960s.

THE BIRTH OF FINANCIAL FUTURES MARKETS

Then, a series of propitious events took place that allowed the CME to create a different kind of futures product. In November 1967, the eminent University of Chicago economist Milton Friedman, believing that the British Pound was overvalued, approached a number of major banks who dealt in the Pound and attempted to sell short the currency. The banks, however, all refused to deal with him, explaining that they only dealt in currencies with their regular commercial customers. Friedman, who as it turned out was correct in his prediction that the British currency would tumble, wrote about his experiences in *Newsweek* and the *Wall Street Journal.* He argued that the Bretton Woods Agreement, which fixed the relative worth of the world's main currencies, should be abandoned and that government restrictions on who could speculate in currencies should be removed. These articles caught the attention of Leo Melamed, a trader and director at the CME. Melamed, who is widely known as the father of the financial futures industry, saw an opportunity for his exchange. He sensed that if currencies were allowed to float freely, their values determined by supply and demand, it would be possible to create currency futures contracts. He also envisioned a world beyond currency futures; one that included metals, interest rates, and equities. Melamed knew that if exchanges could "commoditize" financial products they would herald a revolution in the way that the financial world dealt with risk management. He also knew that the growth potential for financial futures products was far beyond that of grains or meats.

The rest of the world was fed up with Bretton Woods as well, and in August 1971, when the Agreement was abandoned, the CME commissioned Friedman to write a paper entitled *The Need for Futures Markets in Currencies*. Then, Friedman, Melamed, and CME officials pressed their case with the United States government, Federal Reserve, and the retail investment community and banks, both in the United States and overseas. Ten months later, in May 1972, the International Monetary Market of the CME opened for business. In the ensuing years, exchanges introduced additional financial futures contracts in products such as metals, interest rates, and stock indices. Some products failed to attract interest, but others, such as Treasury Bonds, Eurodollars, and the S&P 500 Futures became hugely successful with both commercial users and speculators. The idea of hedging away risk in commodities had evolved to include these new instruments. The world of high finance would never be the same again.

OPEN OUTCRY

Open Outcry refers to the traditional exchange method of bringing together its members to trade a particular product in a defined physical space known as a pit. Customers send orders to the pit and members provide a two-sided market to satisfy the demand. Because all the traders in the pit are competing for the order flow, they scream their bids and offers as loud as possible to attract attention. The theory behind Open Outcry is that competition among exchange members enables customers to execute their orders at a reasonable price. In some pits, hundreds of traders compete. This constant battle for the privilege of making a trade tends to ensure the integrity of the transaction price. This *price discovery* function is the great strength of the Open Outcry system. Critics of Open Outcry bemoan that the U.S. exchanges seem unwilling to abandon the system in favor of electronic trading. Nonetheless, even the critics admit that in many markets the price discovery function is alive and well in the pits.

LIQUIDITY

Liquidity is best understood as *the ability to buy or sell at a reasonable price whenever you need to do so.* While "reasonable price" can mean different things to different people, in our context we can assume that a reasonable price is generally one close to the last traded price. A liquid market is characterized by the presence of at least several well-capitalized market-makers (and in a pit, perhaps hundreds of them); the availability of a two-sided market on demand; and a tight spread between the bid and offer. Without liquidity, a market is doomed to inactivity. In such a situation, buyers and sellers alike will find it virtually impossible to establish a fair price for their assets.

There are two key drivers of the liquidity engine: the *need for immediacy* (also known as a *liquidity event*) and the *supply of immediacy* (Miller, Merton, *Financial Innovations & Market Volatility,* Blackwell, 1991, pp. 24–26). The need for *immediacy* occurs when, for example, a seller determines that it is more sensible to dispose of an asset rather than hold it in portfolio. The price at which the seller is willing to dump the asset is a function of volatility and the risk involved in holding the asset, against the immediate benefit to be appreciated by liquidating. In order to get a price that justifies the sale, the seller looks to the market-maker to provide a reasonable bid. Market-makers, however, do not work for free and are only willing to create immediacy if they are appropriately compensated for their services. They know that in order to show a profit they must cover the direct costs associated with carrying out their transactions (e.g., exchange fees, commissions, cost of membership, salaries and benefits, etc.) as well as the opportunity cost to them of acting in this capacity. Additionally, the price risk, which the seller wishes to eliminate, is transferred to the market-maker. In the end, the assumption of this risk is the central determinant of the two-sided price that the market-maker presents to the seller. When we look at the liquidity model as a whole, it is fair to say that the conditions for an extremely liquid market are present when the demand for immediacy is high and the cost to the market-maker of maintaining a continuous presence is low.

In highly volatile futures markets the demand for immediacy is high. Hedgers are so concerned with price risk that the cost of *not trading* becomes an ongoing concern. Factors such as the use of leverage and lack of diversification in the construction of their portfolios (i.e., a grain hedger rarely has anything but grain in his portfolio) exacerbate this condition. So too does the frequent use of spreading techniques in which, for instance, the December contract is bought and the March sold. These positions entail a significant transaction risk and can expose the user to losses if the market moves adversely prior to the execution of the legs of the spread. Finally, in the commodities markets, commercial users typically build-up or down their portfolios based on seasonal factors. In the financial markets, government auctions, initial public offerings, changes to the composition of the major market indices, and many other factors can cause the commercial user to feel the need for immediacy.

The market-maker is able to service this need effectively because the exchanges have created an environment in which the cost of providing a two-sided market is low. The exchanges create the contracts; underwrite the significant research and development, marketing, clearing, and regulatory costs; and provide the physical infrastructure (i.e., the trading floor and the pits). While the exchange collects a fee for every transaction, the cost is so low relative to the average profit per trade that the market-maker is well rewarded for her efforts. No market is truly liquid all the time, and some markets are more liquid than others. A market such as Eurodollars, for instance, is extremely liquid, while the pork belly market is less so.

ILLIQUIDITY

We have established that in order for liquidity to be present, commercial users must perceive a need for the liquidity, and market-makers must be willing to commit to making a two-sided market. In most cases this model works extraordinarily well. There are, however, occasions where the system breaks down and the market becomes illiquid.

The best analogy to explain this phenomenon is to consider what happens when you go to the bank to withdraw funds during a "run" on the bank. Generally, individual accounts, even large ones, are highly liquid and the bank is able to satisfy all withdrawal requests. This is because there tend to be deposits every day as well as withdrawals and they more or less balance out. The problem arises, however, when depositors question the ability of the bank to live up to its obligations and redeem withdrawal requests. In those situations, it is only natural for the depositors to want to get their money out of the bank as quickly as possible. The ensuing run on the bank ultimately forces it to close, turning depositors into creditors. Ironically in these situations, it is the depositor's very belief in the liquidity of the bank—that one's money is available for withdrawal *at any time*—that ultimately causes its demise due to illiquidity.

Futures markets are subject to similar types of events from time to time. Typically, "runs" on the market will commence with an imbalance of orders in the opening minutes of the day. As the brokers for commercial users flood the pit with sell orders, the market-makers are eventually overcome and adjust their inventories by dumping the contracts they have purchased in the initial selling wave. As the market breaks further, it draws the attention of speculators, such as hedge fund managers and day traders, who see an opportunity to profit by selling short. Technicians, who initiate trades when markets break through support points, enter their sell orders as well. In the case of stock indices, program traders may initiate sell programs, adding to the downward pressure. At some point, just as the bank runs out of money, the market runs out of liquidity.

One way the exchanges and regulators have attempted to address this problem, at least with regard to the U.S. equity markets, is to mandate limits on the amount a market can move lower within a given time period. These limits, known as *circuit breakers,* are set as a percentage of the value of various market indices which, when triggered, force the market participants to stop selling. In the case of large moves, the futures are prohibited from trading lower than the circuit breaker price for a period of time. For example, if the market goes down 2.5 percent it can go no lower for 10 minutes. The market does stay open, however, which allows buyers to come in and force the price higher. If the New York Stock Exchange declares a trading halt in the instance of an even larger move, say 20 percent, the futures must actually shut down until more than 50 percent of the stocks in the Index are reopened. Because the securities, futures, and options exchanges are inextricably linked, when the circuit breakers kick in all of the affected markets follow the same rules. This effectively inhibits anyone from moving the selling pressure from one exchange to another. Interestingly, there is no limit on moves to the upside. While the regulators are apparently fearful of irrational selling pressure, they have no such fears when it comes to irrational buying. In view of market events in the last two years, perhaps a study ought be performed as to whether overly "exuberant" buying is equally damaging to the equilibrium of the marketplace.

Theoretically, one can argue that circuit breakers are unnecessary. In fact, many contend that such attempts at manipulating market dynamics are futile and actually impede the market from righting itself.

The line of reasoning is as follows: Savvy traders seeking a bargain will begin buying when the discount-to-value falls too precipitously. While this theory is popular among University of Chicago economists and other free market purists, regulators, exchanges, and brokerage firms are unmoved by the argument. They fear that in a panic situation confidence in the integrity of the marketplace will be irreparably harmed. Because they would rather not test the consequences of letting the market trade completely unfettered in extreme situations, they have implemented the circuit breakers. The circuit breaker rules have been in effect for a number of years, but it is still unclear whether they work. No one has proven that the rules alleviate systemic pressure and/or encourage buyers to enter the market; after all, buyers might start buying even if there weren't any circuit breakers. More importantly, the fact that the market has yet to experience a cataclysmic fall since the implementation of the rules is hardly proof that it *can't* experience such a fall. Unfortunately, if it happens, we'll only be able to assess the efficacy of the circuit breaker rules as we sort through the wreckage.

THE ROLE OF THE EXCHANGE

Historically, the role of the exchange has been to act as a facilitator of liquidity. Its responsibilities include providing the physical infrastructure and a central gathering place for its members, creating and supporting new futures products, marketing, compliance, clearing, and dealing with regulators. While every market participant has an economic incentive to trade—hedgers want to protect themselves from adverse market movement and market-makers seek to profit by assuming the price risk the hedger wants to avoid—the exchange traditionally has had no such interest. These entities existed as not-for-profit organizations, owned by members and operated for their benefit. Exchanges paid their bills by collecting a fee for every transaction, but since they were not required to produce a profit for shareholders they kept these fees extremely low (fees for members are much lower than for customers, but both groups have appreciated the benefit of the exchange's willingness to provide services at or below break-even). The low fees encouraged trading activity and promoted liquidity, which, in turn, helped the exchange bring in more business. Clearly the relationship between fees and liquidity is of critical importance to the well-being of the organization and all of its users. If fees are raised too high, liquidity may suffer. This can result in the need to charge higher fees which may compromise liquidity further. Exchanges have been extremely careful when determining fees, because as much as they needed cash flow in order to provide value-added services, they could not run the risk of undermining liquidity.

The traditional exchange business model, however, is changing dramatically. Whereas for more than a century the exchanges operated largely as utilities, they have now accepted or are moving towards a for-profit, or *demutualized,* model. The implications for the wider marketplace are considerable. In the traditional model the exchange fulfilled its responsibilities to its members if it provided them with a sufficient opportunity to profit from the trading environment. In a demutualized environment, however, the exchange's primary responsibility is to promote shareholder value. For the first time in their histories, these institutions will be required to cut costs and enhance revenues. It will not happen overnight, but the chief way for them to accomplish these goals will be to transition from the costly Open Outcry system to the efficiencies of electronic trading.

The first futures exchange to experience significant growth was EUREX, and its progress has been a catalyst for many far-reaching changes in the futures industry worldwide. EUREX is a fully electronic exchange at which extremely liquid contracts on European interest rates and stock indices are traded. Created in the 1990s, EUREX grew slowly and then burst into the industry's consciousness in 1998 when it snatched the liquidity away from the LIFFE in its benchmark product, the BUND contract. Almost overnight, market participants switched their allegiance to the EUREX electronic platform and abandoned the costly Open Outcry system employed by the LIFFE in London. This phenomenon was doubly shocking; no exchange had ever before captured the volume in an existing market from another, much less by using technology to accomplish the task. With this event it became clear to everyone that the transition from Open Outcry markets to fully electronic trading had commenced and that an outcome in favor of the PC was inevitable.

In some respects, the critics are right to castigate the U.S. exchanges for their allegiance to the Open Outcry system. It is fair to ask why electronic trading has made only relatively small inroads as compared to situations overseas. It is also fair to ask how customers are advantaged by having to bear the costs of a bifurcated market in which some activity is in the pits and some is online. It is not entirely fair, however, to conclude that the U.S. exchanges have mismanaged the situation because they have lagged behind foreign exchanges with respect to electronic trading. In fact, the exchanges have been working for years to address the extraordinarily difficult choices facing them. They recognize internally that Open Outcry cannot be protected in perpetuity. From a political standpoint, however, because the exchanges are institutions in which members elect other members to run the business affairs of the exchange, exchange leaders have remained largely mute on this subject. They are trying to balance their good business sense against what will be acceptable to the members that can vote them out of office. Accordingly, it should come as no surprise that political considerations will, to some degree, dictate the timeline for change. Furthermore, the Open Outcry system has proven itself remarkably resilient in the face of the changing sentiment towards it. At the time of this writing, Open Outcry exchanges are experiencing record volume in both the pits and their electronic systems.

As if the tug of war between the respective proponents of Open Outcry and electronic trading isn't divisive enough, members of the exchanges have been divided throughout their histories by a myriad of internecine political issues. Members gripe and snipe at each other like Hatfields and McCoys, often unable to work together for the common good. While there are certain interest groups within an exchange that stick together, it is more appropriate to say that there are as many interest groups at an exchange as there are exchange memberships. All members look out for themselves, first and always.

It reminds me of an old joke someone told me once. It seems there were two groups in a synagogue that argued constantly over what the correct custom was for reciting a certain prayer. One group insisted that the custom was to stand during the prayer and the other insisted that the custom was to sit. They could not agree, and so each time the congregation came to the point in the service where this prayer was said, there was an argument. The constant bickering finally became so disruptive that the two groups could not stand it any longer. Someone suggested that representatives of the respective groups should visit Mr. Goldberg. Mr. Goldberg was 104 years old and lived in a nursing home, but as one of the founders of the synagogue, who was there at its inception, he was sure to know which custom was correct.

So the two groups went to the nursing home and explained the problem to the elderly man. They waited for his response, but he said nothing. They thought he didn't hear them so they explained it to him again, but he still said nothing. Finally one group, unable to wait any longer, said, "So Mr. Goldberg, don't you agree that the custom of the synagogue is to stand?" And Mr. Goldberg said, "No, that's not the custom of the synagogue." The other group, of course, was overjoyed. They said, "So Mr. Goldberg, the custom of the synagogue is to sit?" But Mr. Goldberg said, "No. That's not the custom either." At this point, both groups were frantic and cried in unison, "But Mr. Goldberg, we need an answer. Every time we say the prayer there's an argument." At that, Mr. Goldberg looked up, raised a finger to the heavens, and said, "Aha, that's the custom of the synagogue!"

The exchanges are like Mr. Goldberg's synagogue, but instead of two interest groups arguing about what best serves the needs of the organization, depending on the issue there may be 10, 20, or hundreds. So many difficult questions lie ahead: How can the exchanges promote and prioritize shareholder value when the shareholders have such heterogeneous interests? How can they reconcile the interests of members with customers? Small traders with institutions? Pit traders with those who want to trade online? Electronic trading threatens the orthodoxy of the exchanges, the very culture that sustained them throughout the years of their greatest achievements. It is no easy task to break from such cultural bonds. Nonetheless, because it is clear that external factors will eventually force the break—or else threaten the survival of the exchanges—they will choose a middle ground regarding the future of Open Outcry. While they will not reject it overnight, they will move away from it in a very deliberate manner. They have no choice but to proceed slowly, because as Mr. Goldberg might say, choosing the middle ground is the custom of the exchanges.

Here is an example of the kind of intractable problem exchange leaders must try to resolve if they believe that shutting down the Open Outcry system in favor of electronic trading is the best way to promote shareholder value. They must convince the traders in the pits that they are better off in an electronic environment, which is no easy matter. Let's say that there are 50 brokers in a particular pit and the average income per broker is $400,000 per year, or a total of $20 million. That number goes down to zero in a completely electronic environment as there is no longer any need for customers to pay pit brokerage. It will take some pretty skilled exchange salespeople to persuade this group of disintermediated brokers that they are better off because of electronic trading. What will the sales pitch be? "Hey guys, because the exchange can operate more efficiently you'll appreciate an increase in the value of your demutualized shares." Perhaps they should take the enlightened approach: "Hey guys, the exchange will provide you with preferential fees so that as you learn new skills, the learning curve will be less costly." Maybe it requires a "tough love" approach: "Hey guys, you're not guaranteed $400,000 per year in perpetuity. You'll have 100 percent of nothing if we don't make the transition." Perhaps the approach should be a hybrid of all three, but clearly, unless those directly affected by the change believe the amount of benefit will be more than $400,000 per year, they are unlikely to be moved by these or any rationalizations. This example is only one of many potential landmines facing exchange leaders as they step cautiously forward into unmapped territory.

The demutualization process is likely to be a gut-wrenching experience for the exchanges. While it seems self-evident that a properly implemented shareholder value approach would be embraced by shareholders, in the case of the U.S. exchanges, such an approach may instead create enormous distress among the membership. When one considers that the vast majority of exchange revenues derive from clearing fees and that the majority of clearing fees are paid by the members of the exchange, the blueprint for serious conflict becomes apparent. How the exchanges resolve these issues and how quickly they are able to bring about substantive change will ultimately determine whether they can survive.

4
THE EXCHANGES

THE CHICAGO MERCANTILE EXCHANGE, INC. (www.cme.com)

The Chicago Mercantile Exchange, Inc. (CME) is a premier international marketplace for global risk management. Financial institutions, other businesses, and individuals rely upon the CME to trade futures contracts and options on futures, both on the trading floor and on GLOBEX2, the exchange's around-the-clock electronic trading system. The CME has a long history of innovation, which includes becoming one of the first financial exchanges to become a public company (in November 2000). The exchange offers products in four major areas:

1. interest rates
2. stock indexes
3. foreign currencies
4. agricultural commodities

In 2000, more than 231 million contracts with an underlying value of more than $155 trillion changed hands at the CME, and in the first six months of 2001 the exchange was on a pace to eclipse the previous year's volume total by 100 million contracts. Many judge an exchange by the amount of open interest held on its books. By that standard, the CME is the world's leading futures exchange. Its open interest of 12 million positions (as of May 2001) dwarfs that of other futures exchanges, irrespective of whether those exchanges are Open Outcry or electronic.

At the CME, there are three ways to trade futures and options. Some traders buy and sell contracts exclusively using computers, some prefer being face-to-face with other traders on the trading floors, and an increasing number trade both ways.

Who Trades at the CME?

Pension funds, investment advisors, portfolio managers, corporate treasurers, commercial and investment banks, broker-dealers, and individuals all over the world are among those who trade on the exchange as an integral part of their financial management strategy. The right to trade on the floor of the CME is conveyed through Class B shares, which are held by CME members. Members include the world's largest banks and brokerage firms, as well as independent traders and brokers. They are divided into these groups:

- *CME* members (625 B-1 shares) can execute trades in any contract listed on the exchange. On the trading floor, they wear gold badges that show who they are and what they can trade.
- *IMM* stands for International Money Market (813 B-2 shares). IMM members can execute trades in currency, interest rate and stock index futures, as well as futures options (i.e., all IMM and IOM futures and futures options). IMM badges are green.
- *IOM* means Index and Option Market (1,287 B-3 shares). IOM members can execute trades in index futures contracts, random length lumber contracts, and all option futures. IOM badges are blue.
- *GEM* is the Growth and Emerging Markets Division (467 authorized B-4 shares). GEM members may execute trades in a number of products, including those related to emerging market countries. GEM badges are black or brown.

The CME as a Public Company

On November 13, 2000, the CME finalized its transformation from a not-for-profit, member-owned organization to a for-profit, shareholder-owned corporation. Chicago Mercantile Exchange, Inc. is a public company incorporated in Delaware. It is also the first U.S. exchange to demutualize by converting its membership interests into shares of common stock that can trade separately from exchange trading privileges.

In demutualizing, the exchange's goals included unbundling members' equity value, providing currency for working with strategic partners, and supporting the exchange's expansion by giving it access to the capital markets. Demutualization encourages a greater level of fiscal discipline, enables the exchange to more aggressively pursue new business opportunities, and helps solidify the CME's position as a premier global marketplace.

As a result of the CME's demutualization, nearly 26 million shares of Class A common stock were allocated on a 3-2-1 basis to 3,200 members of the CME, IMM, and IOM divisions. These Class A shares have the traditional features of common stock. In addition, about 5,000 shares of Class B common stock were issued to equity members in a series corresponding to the former membership divisions. Each CME member received a B-1 share, each IMM member received a B-2 share, each IOM member received a B-3 share, and each GEM member received a B-4 share. Each B series share confers the trading privileges associated with the membership interests that are converted into each series, along with the traditional features of common stock. Class B shares are bought, sold, and leased through the CME's Membership Department. The CME announced in April of 2001 that it was planning an initial public offering and it is likely that by the end of the first quarter of 2002, you will be able to trade shares of CME, Inc.

A History of Innovation

The Chicago Butter and Egg Board was founded in 1898 and evolved into the Chicago Mercantile Exchange in 1919. At that time, futures were offered only on inanimate agricultural products, such as butter and eggs. Since that time, the CME has pioneered a number of innovative products, services, and systems. These include:

- *1961*—CME introduces a frozen pork belly futures contract—the first futures contract based on frozen, stored meats. (Pork bellies are used to make bacon.)
- *1964*—A live cattle futures contract begins trading at CME. This is the first futures contract to be based on a non-storable commodity.

- *1972*—CME creates the world's first financial futures contracts by introducing futures on seven foreign currencies.
- *1981*—Eurodollar futures begin trading, becoming the first contracts to be settled in cash, rather than the physical delivery of a commodity. The Eurodollar futures contract is now the world's premier short-term interest rate product.
- *1982*—CME introduces stock index futures products, including the first successful equity index contract, the Standard & Poor's (S&P) 500 Index futures.
- *1984*—The first international link between futures exchanges is established: A mutual offset trading link between CME and Singapore Exchange Derivatives Trading Ltd. (SGX).
- *1992*—GLOBEX®, the world's first international post-market electronic transaction system, begins live trading on June 25. Its successor, GLOBEX2, is introduced in 1998.
- *1997*—E-mini S&P 500 contracts—electronically traded products that are one-fifth the size of the standard S&P 500 Index futures—are introduced. They soon become one of CME's most successful products, paving the way for a new e-mini product line.
- *2000*—CME becomes the first U.S. financial exchange to demutualize by converting its membership interests into shares of common stock that can trade separately from exchange trading privileges.

With customers all over the world, a global product line, around-the-clock electronic trading, and strategic alliances with other exchanges, the CME is truly a global marketplace offering 24-hour access to its products. Various business alliances will soon give CME customers and members access to products on eight exchanges in all the world's major time zones:

- Under the GLOBEX Alliance, members of CME, ParisBourse[SBF] SA, Singapore Exchange Derivatives Trading Ltd. (SGX), Brazil's Bolsa de Mercadorias & Futuros, the Montreal Exchange, and Spain's MEFF will benefit from trading privileges and direct access to the electronically traded products of all the alliance markets through one single technical access point, along with cross-margining of positions in order to reduce the capital requirements of their customers and members.

- An agreement with the London International Financial Futures and Options Exchange (LIFFE) provides cross-access to the two exchanges' respective electronically traded products and cross-margining of the world's most actively traded short-term interest rate products.
- Under a mutual offset system with the Singapore Stock Exchange (SGX)—in place since 1984—market participants can establish a position in certain products at CME and offset it at the SGX, and vice versa.
- In 2000, the CME announced an alliance with the Tokyo Stock Exchange. The goal of the alliance is to further develop the two exchanges' fixed income and equity derivatives markets through various initiatives, such as interconnecting the exchange's electronic trading systems.

B2B Marketplaces

The CME also works with new types of marketplaces, such as the rapidly growing online business to business (B2B) arena. These marketplaces use the Internet to offer a range of services or products, linking companies in a single industry or with an interest in a particular type of product. For example, a B2B marketplace might allow companies to view catalogs, negotiate prices, and order supplies online. In 2000, the CME announced plans to work with CheMatch.com, Inc. to jointly develop and market a co-branded complex of certain chemical futures and options products. This represents the first joint development project between a futures exchange and a B2B marketplace to create risk management products targeted to a specific industry, as well as the first time futures products will trade on a futures exchange with an electronic link to an online B2B marketplace.

The CME's Financial Safeguards

Institutions and individuals turn to CME products to help manage their risk. Through a financial safeguard and "clearing" system unique to futures markets, the exchange ensures that the counterparties to each trade will make good on the contract. The CME's Clearing House guarantees each and every trade, whether made on the trading floor or through GLOBEX2. The Clearing House does this by acting as a buyer to every sell order and a seller to every buy order. In addition, all CME members are qualified (guaranteed) by a CME clearing member firm.

Before making any trade, members and customers must post a minimum performance bond or margin, the "good faith" or earnest money that CME requires before traders can take a position in CME's futures and options markets. The CME developed the Standard Portfolio Analysis of Risk (SPAN®) system for calculating performance bond requirements for each product based on potential losses that could occur. SPAN is now the margin system of choice in financial capitals worldwide, including Chicago, New York, London, Paris, Oslo, Singapore, Hong Kong, Sydney, Tokyo, and Osaka.

The CME's Clearing House collects performance bond money from its members and customers twice daily, based on the firm's trades during the day and the positions still outstanding after the market's close. This activity is known as "mark-to-market," and significantly reduces the chance of default. As one measure of the success of the CME's financial safeguards, there has never been a clearing member firm default at the exchange in its more-than-100-year history.

On an average day, the Clearing House acts as custodian for $30 billion in performance bond assets that back the market positions taken by customers, members, and member firms. To better manage the clearing process, the CME developed the CLEARING 21® system with the New York Mercantile Exchange (NYMEX) in 1994. Widely considered the most efficient and powerful clearing system in the derivatives industry, CLEARING 21 is being adopted by exchanges around the world.

GLOBEX2 (G2)

GLOBEX2 is the CME's electronic trading system that matches all of the exchange's electronic trades and provides CME-generated market data. As of May 2001, G2 handles nearly 200 transactions per section (TPS) and by 2003 is expected to handle 500 TPS, a transaction is defined as any message sent to the trade matching engine, so it does not necessarily have to result in a trade (e.g., a bid/offer or canceling a bid/offer is a transaction). The G2 trade matching engine is extremely stable with uptime availability at 99.999 percent. This means the trade matching engine is available, on average, for all but two minutes per month. The trade matching engine is designed for speed and thrives on it. As trading volume increases, the engine runs faster, kind of like a sports car which functions better the harder it is pushed. Accordingly, the response time, on average, is less than one second.

Current G2 functionality includes:

- Anonymity in the marketplace
- Order entry and management
- Trade cancellation
- Market depth (the "book")
- FIX/API Interface (allows Independent Software Vendors to write software that interfaces with the G2 trade matching engine)
- Request for Quote function
- Audit Trail logging
- Customer support through the GLOBEX Control Center (GCC)

There are two avenues for access to the G2 matching engine; either CME-offered access, or access through customized third party interfaces.

Avenue 1: CME-Offered Access. Called GLWIN (GL for the company that developed and designed it, and WIN for Windows), this is a proprietary system that was the *first* system developed for access to the G2 host engine. The screen displays what people have come to think of as the traditional "Globex" screen:

- Provides access to CME products and CME market data
- Can provide optional access to LIFFE

There are currently three options for obtaining CME-offered access:

Option 1: Virtual Private Network (VPN)

- Newest way to connect to CME-offered access
- Customers trade via a secure connection to the Internet
- Uses GLWIN software (the first GLOBEX interface)
- Intended for low-volume, single-connection users

- Customers responsible for the circuitry (the "wire" that connects them to the GLOBEX engine); users decide if they want a cable, DSL, or T1 connection
- The "lowest tech" option available
- Currently the quickest way to get GLOBEX2 access (connection usually possible within three to five days)
- Currently the lowest cost form of access ($400/month to the CME for the license fee, and communication costs that can be as low as $50/month)

Option 2: CME-offered software and circuitry

- Customer uses own computer
- CME provides GLWIN software
- CME manages cable installation directly to the GLOBEX engine
- Bandwidth choices available at different costs

Option 3: CME-offered hardware (terminal), software, and circuitry

- Earliest option developed
- Often referred to as the Managed Workstation option
- CME provides hardware (often called a GLOBEX terminal)
- CME installs GLWIN software
- CME manages the cable circuitry connection to the GLOBEX engine
- High-end cost ($1,400 per month)

Avenue 2: Custom, Third-Party Interfaces.

- Front-end trading systems developed by firms and Independent Software Vendors that are approved and certified by the CME
- Customized and customizable to firm/ISV preferences
- Development of additional interfaces encouraged by the CME, to better serve the needs and goals of firms and developers
- Connection to G2 via FIX order routing systems

    (FIX stands for Financial Information Exchange specifications, which are industry-developed, flexible software formats that accommodate various types of trading)

Two kinds of FIX interfaces:

FIX 2.3
- Supports all G2 order types
- Supports optional order management systems

2.3 Express
- Faster access to GLOBEX because this system supports fewer types of order types (no Market If Touched or Stop Orders)

The Future of G2: iLink

Like Avenue 2, iLink access begins with customized front-end systems developed by third parties. Access to the G2 trading engine is facilitated by state-of-the-art "middleware" software, provided by a company called TIBCO, that optimizes all system interactions. The goal of this approach is to provide a single, broad "gateway" for access to all the exchange's major trade-related functions.

Direction of Change for the iLink Gateway System now being developed:

- From exchange-based, legacy/proprietary systems to standardized industry protocols
- Toward greater speed, volume, and scale
- Toward more *customer-friendly* or *customer-facing* systems and away from more exchange-oriented systems
- Toward a single electronic gateway or "expressway" for all basic exchange functions (i.e., order entry/matching, market data) and clearing from separate gateways for each function
- Toward greater *fault tolerance*—delivery of uninterrupted service despite any failed component
- Toward improved *scalability*—ability to handle significantly larger volumes efficiently, and not only to function well in the rescaled situation, but to take full advantage of it.
- Toward higher *performance*—faster response time per trade as well as faster throughput time for all daily transactions
- Toward improved *manageability*—increased cost efficiency and effectiveness in terms of resources, systems, and people

- Toward greater *simplicity*—designed so that a minimum number of moving parts are involved in solving any problem
- Toward more *nimbleness*—designed for easy re-organization and re-assembling to accommodate different business models

CME Products and First Date of Trading

Agricultural   Feeder Cattle (futures 11/30/71, options 1/9/87)
E-mini Feeder Cattle (futures 9/19/00)
Frozen Pork Bellies (futures 9/18/61, options 10/3/86)
Live Cattle (futures 11/30/64, options 10/30/84)
Lean Hogs (futures 11/10/95, options 11/13/95)
E-mini Lean Hogs (futures 7/25/00)

Currencies   Australian dollar (futures 1/13/87, options 1/11/88)
Brazilian real (futures & options 11/8/95)
British pound (futures 5/16/72, options 2/25/85)
Canadian dollar (futures 5/16/72, options 6/18/86)
Euro FX (futures & options 1/4/99)
E-mini Euro FX futures (10/7/99)
E-mini Japanese yen futures (10/7/99)
Japanese yen (futures 5/16/72, options 3/5/86)
Mexican peso (futures & options 4/25/95)
New Zealand dollar (futures & options 5/7/97)
South African Rand (futures 5/7/97)
Swiss franc (futures 5/16/72, options 2/25/85)

Interest Rates   Three-month Eurodollar (futures 9/9/81, options 3/20/85)
Euroyen (futures 3/6/96)
Euroyen LIBOR (futures 4/1/99)
Ninety-day U.S. Treasury Bill (futures 1/6/76, options 4/10/86)
One-year U.S. Treasury Bill (futures 3/28/94)

One-month Federal Funds Rate (futures 10/12/95)
One-month LIBOR (futures 4/5/90, options 6/12/91)
Ten-year Japanese Government Bond (JGB) (futures 1/21/99)

Indexes Standard & Poor's 500 (futures 4/21/82, options 1/23/83)
E-mini S&P 500 (futures & options 9/9/97)
S&P 500/BARRA Growth Index (futures & options 11/6/95)
S&P 500/BARRA Value Index (futures & options 11/6/95)
Standard & Poor's MidCap 400 (futures & options 2/13/92)
NASDAQ-100 Index® (futures & options 4/10/96)
E-mini NASDAQ-100 Index® (futures 6/21/99)
Russell 2000 Stock Price Index (futures & options 2/4/93)
FORTUNE e-50 Index™ (futures 9/5/00)
Nikkei 225 (futures & options 9/25/90)
GSCI$^{TM/SM}$ (futures & options 7/28/92)

Product	Pit Trading Hours	Globex Hours M-Th	Globex Sun/Hol
Live Cattle	9:05 am-1:00 pm	None	None
Feeder Cattle	9:05 am-1:00 pm	None	None
E-mini Feeder Cattle	9:05 am-1:00 pm	9:05 am-1:00 pm	None
Lean Hogs	9:10 am-1:00 pm	9:10 am-1:00 pm	None
E-mini Lean Hogs	9:10 am-1:00 pm	9:10 am-1:00 pm	None
Currencies	7:20 am-2:00 pm	4:30 pm-4:00 pm	5:30 pm-4:00 pm
Interest Rates	7:20 am-2:00 pm	2:10 pm-7:05 am	5:30 pm-7:05 am
S&P 500	8:30 am-3:15 pm		5:30 pm-8:15 am
E-mini S&P 500		3:45 pm-3:15 pm	5:30 pm-8:15 am
NASDAQ 100	8:30 am-3:15 pm		5:30 pm-8:15 am
E-mini NASDAQ 100		3:45 pm-3:45 pm	5:30 pm-8:15 am

Exhibit 4-1 CME Pricing Schedule

	Equity Indices	E-minis	Currencies	Agricultural
Clearing Fees Per Round Turn				
Equity Member	$0.30	$0.15	$0.10	$0.18
Lessee Member	$0.76	$0.38	$0.50	$0.58
106 H/J/N Firm	$1.16	$0.58	$0.90	$0.98
CBOE Member	$1.36	$0.68	N/A	N/A
Non-Member	$1.56	$0.78	$1.20	$1.28
GLOBEX2 Fees Per Round Turn				
Equity Member	$1.00	$0.50	$1.00	$1.00
Lessee Member	$1.00	$0.50	$1.00	$1.00
106 H/J/N Firm	$1.00	$0.50	$1.00	$1.00
CBOE Member	$3.00	$1.50	N/A	N/A
Non-Member	$3.00	$1.50	$2.00	$3.00

Interest Rate Products

Clearing Fees—First 3,750 Round Turns
Equity Member	$0.08
Lessee Member	$0.29
106 H/J/N Firm	$0.49
Non-Member	$0.64

Clearing Fees—3,751 to 7,500 Round Turns
Equity Member	$0.03
Lessee Member	$0.24
106 H/J/N Firm	$0.44
Non-Member	$0.59

Clearing Fees—less than 7,500 Round Turns
Equity Member	$0.01
Lessee Member	$0.22
106 H/J/N Firm	$0.42
Non-Member	$0.57

GLOBEX2 Fees
Equity Member	$0.50
Lessee Member	$0.50
106 H/J/N Firm	$0.50
Non-Member	$1.60

THE CHICAGO BOARD OF TRADE (www.cbot.com)

The Chicago Board of Trade (CBOT®) offers two venues for trading: the Open Outcry pits and the electronic trading platform—a/c/e^{SM}. The a/c/e trading platform, which is the result of a business and technology partnership between the Chicago Board of Trade and EUREX—two of the world's leading derivatives exchanges—provides traders access to the same financial, equity, and agricultural contracts that are traded in the CBOT Open Outcry pits. But the electronic markets accessed on a/c/e bridge global time zones, offering liquid markets for nearly 24 hours a day, making it possible for traders to use these markets during their normal business hours, anywhere in the world.

Because direct access to trading CBOT products is restricted to exchange memberships, market users who are not members and want to trade CBOT products electronically must establish a relationship with a CBOT clearing firm. The clearing firm can offer a variety of methods for obtaining electronic access to CBOT products, ranging from the Internet (to route orders through the clearing firm), to the firm's proprietary software, and to the firm's own wide area network (WAN). Depending on the options available through the clearing firm, the cost of access can be as economical as going through the firm's website where market participants use their own PCs, or as "full service" as leasing work space with a workstation and quote screens at one of the CBOT clearing firm's offices.

A direct connection means that the individual or firm trades from a workstation that is connected to a clearing firm's a/c/e server. In order for individuals or trading firms to obtain direct access to CBOT products, they must first meet two requirements:

1. Confirm access to an existing CBOT membership or purchase a CBOT membership.
2. Establish a business relationship with a CBOT clearing member firm. The clearing firm must approve the method of connecting to the a/c/e platform, including the option of connecting directly to the clearing firm's a/c/e servers. The list of CBOT clearing firms and contacts can be found on the CBOT web site at *www.cbot.com*.

The extent to which one can trade CBOT products is based on the type of membership: owned or leased. A CBOT Full Membership can trade all products, whereas a CBOT Associate Membership is restricted to financial products. For the current bid, offer, and last sale for all CBOT membership types, visit the CBOT website.

All transactions on a/c/e incur an exchange transaction fee and a clearing fee in addition to any access fees charged by the clearing firm. Member fees are significantly lower than non-member fees. All trades executed on a/c/e are cleared through the Board of Trade Clearing Corporation (BOTCC®) and are completely fungible with contracts traded in the Open Outcry pits.

For the software windows that traders use to enter and monitor orders and markets, a wide variety of offerings exist, since the a/c/e platform incorporates an open interface with a standardized API. Traders may choose from among the basic functionality for trading provided with the a/c/e software's Graphical User Interface, the rich and specialized functionality developed by one of the numerous Independent Software Vendors, or a proprietary interface developed by one of the financial firms.

CBOT Products Traded on a/c/e

Financial Products

U.S. T-bond futures
Options on U.S. T-bond futures
10-year U.S. T-note futures
Options on 10-year U.S. T-note futures
5-year U.S. T-note futures
Options on 5-year U.S. T-note futures
2-year U.S. T-note futures
Options on 2-year U.S. T-note futures
CBOT 10-year Agency Note futures
Options on CBOT 10-year Agency Note, futures
CBOT 5-year Agency Note futures
Options on CBOT 5-year Agency Note, futures
Mortgage futures
Options on Mortgage futures
30-day Fed Funds futures

Long-Term Municipal Bond Index futures
Options on Long-Term Municipal Bond Index futures

Commodities

Options on Soybean futures
Corn futures
Options on Corn futures
Oat futures
Options on Oat futures
Rough Rice futures
Options on Rough Rice futures
Soybean futures
Soybean Meal futures
Options on Soybean Meal futures
Wheat futures
Options on Wheat futures
100-ounce Gold futures
5,000-ounce Silver futures

Index Products

CBOT Dow Jones^{SM} Industrial Average future
Options on CBOT Dow Jones^{SM} Industrial futures
CBOT Dow Jones^{SM} Composite Average futures
CBOT Dow Jones^{SM} Transportation Average futures
CBOT Dow Jones^{SM} Utility Average futures

Trading Hours for CBOT Products on a/c/e

a/c/e Platform (Sunday-Friday)	CBOT Hours (Chicago Time)	London	Frankfurt	Tokyo (During Daylight Savings)*	Tokyo (During Non-Daylight Savings)*
Financial	8:00 pm - 4:00 pm	2:00 am - 10:00 pm	3:00 am - 11:00 pm	10:00 am - 6:00 am	11:00 am - 7:00 am
Equity	8:15 pm - 4:00 pm	2:15 am - 10:00 pm	3:15 am - 11:00 pm	10:15 am - 6:00 am	11:15 am - 7:00 am
Metals	8:15 pm - 4:00 pm	2:15 am - 10:00 pm	3:15 am - 11:00 pm	10:15 am - 6:00 am	11:15 am - 7:00 am
Agricultural	8:30 pm - 6:00 pm	2:30 am - 12:00 pm	3:30 am - 11:00 pm	10:30 am - 8:00 pm	11:30 am - 9:00 pm

*Chicago Daylight Savings starts on the first Sunday in April and ends on the last Sunday in October. London and Frankfurt Daylight Savings starts on the last Sunday in March and ends on the last Sunday in October. Tokyo does not observe Daylight Savings.

Exhibit 4-2 CBOT Pricing Schedule

> Open Outcry

Chicago Board of Trade
Exchange Transaction Fees (Broker Fee's)
Fee Schedule (on a per side basis)
EFFECTIVE SEPTEMBER 1, 2000

	*Full & AM	*Professional CBOT	*Interest HOLDERS	*Other CBOT
Futures	0.05	0.05	0.05	0.05
Deliveries	0.05	0.05	0.05	0.05
Cash Exchanges (EFP's)	0.05	0.05	0.05	0.05
Transfers	N/A	N/A	N/A	N/A
Options	0.05	0.05	0.05	0.05
Assignments	0.05	0.05	0.05	0.05
Exercises	0.05	0.05	0.05	0.05
Expirations	0.05	0.05	0.05	0.05
Give-Up Transactions To	0.05	0.05	0.05	0.05
APS	0.05	0.05	0.05	0.05
Cross Exchange Fees	N/A	N/A	N/A	N/A

> Electronic Trading

Alliance Fees (a/c/e)
Exchange Transaction Fees
Fee Schedule (on a per side basis)
EFFECTIVE AUGUST 27, 2000

	Financial Member	Financal Delegates	Financial Non-Member	Ag's, Metals Indexes Member
Futures	0.25	0.45	0.8	0.25
Deliveries	0.25	0.45	0.8	0.25
Cash Exchanges (EFP's)	0.25	0.45	0.8	0.25
Transfers	N/A	N/A	N/A	N/A
Options	0.25	0.45	0.8	0.25
Assignments	0.25	0.45	0.8	0.25
Exercises	0.25	0.45	0.8	0.25
Expirations	0.25	0.45	0.8	0.25
Give-Up Transactions To	0.25	0.45	0.8	0.25
APS	0.25	0.45	0.8	0.25
Cross Exchange Fees	N/A	N/A	N/A	N/A
Calendar Spreads	0.25	0.45	0.8	0.25

Transaction fees are as defined by Rule 450.00 of the CBOT Rules & Regulations.
For NFA information, 781-1382.
For Clearing information, 786-5700.
*Subject to CFTC Review.

Delegate CBOT®	*Non-Member Licensed Contracts CBOT®	All Other Contracts Non-Member CBOT®	MIDAM	Non-Member MIDAM
0.2	0.75	0.5	0.02	0.25
0.2	0.75	0.5	0.02	0.25
0.2	0.75	0.5	0.02	0.25
N/A	N/A	N/A	N/A	N/A
0.2	0.75	0.5	0.02	0.25
0.2	0.75	0.5	0.02	0.25
0.2	0.75	0.5	0.02	0.25
0.2	0.75	0.5	0.02	0.25
0.2	0.75	0.5	0.02	0.25
0.2	0.75	0.5	0.02	0.25
N/A	N/A	N/A	N/A	N/A

Ag's, Metals Indexes Delegates	Ag's, Metals Indexes Non-Member
0.75	1.50
0.75	1.50
0.75	1.50
N/A	N/A
0.75	1.50
0.75	1.50
0.75	1.50
0.75	1.50
0.75	1.50
0.75	1.50
N/A	N/A
0.75	1.50

THE SYDNEY FUTURES EXCHANGE[1] (www.sfe.com.au)

For most of its history, Sydney Futures Exchange (SFE) has focused on the trading and clearing of futures and options contracts for the Australian and New Zealand marketplaces. It is now evolving beyond its futures origins by broadening its range of products and services, and repositioning itself into a more significant financial services provider to the Asia Pacific region.

Since November 1999, SFE has recorded a number of major achievements:

- It became fully electronic by closing its trading floor and moving to around-the-clock screen trading.
- It demutualized and incorporated, separating ownership of the exchange from the right to access its markets.
- It introduced a new corporate governance structure, including the appointment of senior executives from the wider financial community with a diversity of trading, banking, and public policy experience to the Board.
- It listed its shares on the exempt market conducted by Austock with a view to migrating to an ASX listing in 2002.
- It merged its clearing house with Austraclear Limited to create a centralized clearing and settlement service provider for the clearing of both derivatives and over-the-counter (OTC) debt products in Australia and New Zealand.

These achievements have given SFE the corporate flexibility and diversified revenue base to position itself as a provider of a broad range of financial services. The SFE's transformation is not change for change's sake. The reinvention is part of a deliberate strategy that recognizes the changes taking place in financial markets globally. The SFE has taken the position that derivatives exchanges need to develop and enhance their core competencies in areas over and above execution; that markets are converging and execution has become a largely commoditized activity. Reliance on exchange-traded derivatives products and their turnover

---

[1] Based on remarks made in March of 2001 by Robert Elstone, CEO of the SFE.

volumes alone is an unwise business model. The failure rate among newly listed derivatives products globally is over 90 percent, and exchange volumes depend on the unpredictable volatility of underlying markets; hence, the imperative of diversifying revenue sources.

The SFE's determination to seek a future beyond futures was illustrated by its merger with Austraclear in December 2000. Austraclear is the central securities depository, clearing and settlement organization for all Australian dollar denominated (non-Commonwealth Government) debt securities. It holds on average $A280 billion worth of securities in safe custody and settles more than $A6.2 billion in securities on a daily basis. The merger creates a one-stop-clearing-shop for both derivatives and OTC debt products across Australia. It has also created the capacity to perform a depository and registry function, as well as to offer both central counterparty and delivery versus payment clearing services to customers. In simple terms, SFE's merger with Austraclear provides an immediate source of service and revenue diversification away from a dependence on futures trading volumes.

The intellectual capital and expertise the SFE has gained in relation to clearing provides it with a significant opportunity to apply these skills to non-futures products in a variety of areas. In addition to the Austraclear-related benefits, SFE has moved to add value for bond and repo participants in the cash markets by introducing futures-style "novation" clearing. The introduction of bond and repo clearing is the first OTC central counterparty clearing facility offered to the Australian market. The benefits to users include minimized counterparty risk exposure, multi-lateral balance sheet netting efficiencies, and potential margin offsets with SFE futures and options portfolios. There is no reason why successful exchanges such as the SFE, which have the expertise to add value from the front to the back-end in areas such as contract design, regulation, and clearing cannot also apply this expertise to non-futures markets. This is part of the SFE's vision.

The SFE was one of the first of the world's important futures exchanges to close down its trading floor and make the transition to electronic trading. It realized earlier than most of its counterpart exchanges in other parts of the world that the case for converting to screens was incontrovertible. The SFE is now open around-the-clock and the traditional barrier of member-only entry to its markets has been removed. Market liquidity and depth have been enhanced accordingly.

The SFE's achievements since November 1999 have coincided with the embrace of strategic objectives appropriate for a commercially focused, shareholder-owned corporation competing on the international stage. These include:

- Strengthening its position as the major exchange for the trading of $A/$NZ derivatives products and for the clearing of both exchange-traded and OTC financial products
- Distributing its products more widely, both domestically and internationally, by extending access to new and existing users
- Expanding its markets through the development of new products and by participating in international alliances aimed at providing user benefits in the areas of cost, liquidity, and capital
- Diversifying into new markets by developing non-derivatives trading and clearing services

The SFE is also committed to managing its capital base in a manner that creates shareholder value, pursuing cost and revenue accountability in order to identify opportunities for revenue growth and cost reductions, and aligning the incentives of management with those of shareholder-return objectives. In this way, the SFE is applying business imperatives and subjecting its activities to robust analysis in a manner that it has never done before.

In recent months, the SFE has undertaken a number of high profile activities to begin meeting many of its objectives. Alliances have an important part to play in the SFE's strategy, as long as they are predicated on user-driven demand for trading and/or clearing opportunities. The SFE entered into such an alliance in January 2001 with the Hong Kong Exchanges and Clearing Limited (HKEx). The strategic partnership is designed to develop a range of new derivative trading and clearing services for the Asia Pacific exchange-traded and OTC marketplaces. The specific areas of cooperation targeted by the alliance are cross-access arrangements, omnibus account clearing, and reduced operating costs through the sharing of global networks and market operation infrastructure. The international reputation, geographical location, and product offering of the SFE and HKEx make them natural business partners.

On the product front, the SFE has devoted considerable effort to enhancing its core product range and rationalizing those areas with inadequate commercial returns. A new product was launched in February 2001, based on the world's largest financial market—foreign exchange. The $A/$U.S. currency futures contract has the market-maker support of Deutsche Bank and has been trading, on average, almost 400 contracts a day.

The $A/$U.S. futures contract provides a means of access to the foreign exchange market for private and professional traders alike. If it continues to trade healthily, it is likely to be the first in a suite of new currency products. The SFE is confident of the $A/$U.S. futures contract's success, despite the dubious record of new futures products. Among the reasons for this confidence is the fact that the SFE has the support of one of the world's largest foreign exchange banks—an arrangement it wishes to continue with other new product listings. The SFE believes the seeking out of co-sponsors for market initiatives will become a popular and commercially necessary trend.

With regard to technology, the SFE is expecting to implement OM SECUR technology as the cornerstone for improved clearing services in the third quarter of 2001. This benchmark technology will allow the exchange's clearing house to offer web-based reporting systems, flexible access, and multi-currency capabilities to its clearing customers. Such an upgrade in relation to clearing complements enhancements recently made to the SFE's electronic trading system, SYCOM®. These enhancements include expanded distribution options via the improved functionality and capacity of the Interface (the gateway to connect into SFE), the ability to view and trade, market depth, fully interactive spread trading, increased transaction through-put, and industry standard matching and order types. These enhancements provide SYCOM users with a state-of-the-art open architecture trading system.

The expanded distribution options available through SYCOM are evidenced by the SFE's growing network of communication hubs in the world's financial centers. It is through these hubs that the exchange provides customers around the world with the opportunity to directly access SFE markets via the SYCOM Interface. Communication hubs are already operational in London, Tokyo, Hong Kong, and New Zealand. Similar hubs are planned to go "live" in New York and Chicago by the third quarter of 2001. Up to 18 percent of SFE's traded volume is already generated offshore; the installation of hubs in these two key financial centers will push that percentage higher and increase

SFE's overall trading volume. The Asian time zone is one factor that bodes well for the growth of the SFE. Because Sydney's morning coincides with late afternoon in New York and Chicago, it is only natural that some traders will simply continue trading SFE products when the U.S. markets' day session winds down. For instance, U.S. bond traders may be interested in using their interest rate expertise to trade the SFE's 3-Year Bond futures contract, one of the ten most heavily traded government bond contracts in the world. In fact, in late May 2001, the SFE was enjoying one of its strongest trading periods ever in its bonds. On a single day alone, the 3-Year Bond traded 207,319 contracts—a new daily record—and total exchange volume amounted to 380,172 futures and options contracts traded—the highest trading day in almost three years.

Looking ahead, the SFE has announced its intention to migrate from the Austock exempt market to a full listing on the ASX toward the end of the first quarter of 2002. Although the exempt market has operated satisfactorily overall, with the share price rising since listing, the SFE is mindful of the desirability from a shareholder perspective of migrating to a more liquid market. A move to the ASX at that time makes sense because the financial results for SFE's first full-year of operations as a commercially focused, shareholder-owned entity will be available. A listing on the ASX will also symbolically represent the culmination of SFE's corporate transformation.

EUREX (www.eurexchange.com)

EUREX, the European Exchange, is the world's largest completely computerized derivatives exchange. EUREX broke new worldwide trading records in 2000 when its annual volume exceeded 454 million contracts. The exchange continued to set new records in 2001 with more than 163 million contracts traded in the first quarter alone.

EUREX was created in 1998 through the merger between the former Deutsche Terminbörse (DTB) and the Swiss Options and Futures Exchange (SOFFEX). In contrast to Open Outcry exchanges, trading at EUREX is not conducted at a central location but via an electronic network. The exchange is dedicated to providing its members with cost-effective access to a global liquidity network, modern clearing facilities, and an innovative product line.

EUREX currently lists options on German, Swiss, Finnish, Dutch, French, and Italian blue chip stocks—futures and options based on German, Swiss, Finnish, and European indices and interest rate products that cover the German yield curve from one month to thirty years, and the Swiss yield curve from eight to thirteen years. EUREX's benchmark product, the Euro-BUND future, is also the world's most heavily traded derivatives contract with more than 48 million contracts traded during the first three months of 2001.

The EUREX system is order-driven and all quotes (defined as two-sided orders) entered by market-makers are matched using the same principle as the orders entered by non-market-makers in the system. Index and interest rate products traded on EUREX are matched using the "Price-time-Priority" principle. Orders are sorted by type, price, and entry time, with market orders having the highest priority. Money market products are matched according to the Pro-Rata matching principle, where priority is based solely on price. The system constantly updates the ten best bids and offers and makes this information available to all participants.

Participation at EUREX can be direct or indirect. Direct participation is possible by becoming a General Clearing Member (GCM), Direct Clearing Member (DCM), or a Non-Clearing Member (NCM) at EUREX. The difference between these memberships is their role in the clearing process. NCMs may execute trades for their own account or for their customers' accounts. They cannot clear and therefore require a clearing agreement with an EUREX GCM or DCM. Both GCMs and DCMs are allowed to clear their own trades, those of their customers as well as business for their affiliates. In addition, a GCM may clear trades for an NCM. Membership status is separate from ownership and there is no market for buying and selling EUREX memberships. Indirect participation is possible by becoming a customer of an existing member.

Market participants terminals are linked via dedicated lines, the Internet, or a combination of the two to Access Points located in cities in Europe, the United States, and the Asia-Pacific region. These Access Points are directly connected to the host computer in Frankfurt. There is no fee for participation as an NCM; however, all participants must pay a minimum transaction fee to the exchange depending on the access connection type that they have chosen.

The EUREX trading system and its network are also used by EUREX to develop new markets and exchanges such as the EEX—the European Energy Exchange (a regulated Exchange for energy contracts) and EUREX Bonds (an ECN for European cash bonds). EUREX also rolled out its trading system to a/c/e (alliance/cbot/EUREX) in August 2000, allowing CBOT products to be traded using the same worldwide network and matching engine technology.

Key EUREX Futures Products and Clearing Fees

DAX-30 German Stock Index (Euro 25 per DAX Index point)	1.00 EUR/Round Turn
DJ Euro-Stoxx 50 Index (Euro 10.00 X Index)	.60 EUR/Round Turn
EURO-SCHATZ: 2-Year German Government Bond	.40 EUR/Round Turn
EURO-BOBL: 5-Year German Government Bond	.40 EUR/Round Turn
EURO-BUND: 10-Year German Government Bond	.40 EUR/Round Turn

LIFFE (www.liffe.com)

The London International Financial Futures Exchange (LIFFE) is a completely electronic exchange, based in the United Kingdom. In May 1998, with the overwhelming support of its members, LIFFE began development of an electronic trading platform capable of replacing floor trading. The new system, called LIFFE CONNECT™, was designed by LIFFE in conjunction with customers and selected Independent Software Vendors (ISVs) to handle all LIFFE's financial futures and options contracts, the first of which, Individual Equity Options, were migrated off the floor and onto the screens on November 30, 1998. Since then, subsequent contract migrations/listings onto LIFFE CONNECT have been conducted on a phased basis. At the heart of LIFFE CONNECT is the LIFFE Trading Host, where all orders are received and matched. The Host also performs price reporting and dissemination, displaying all transacted prices together with the aggregate size of all bids and offers above and below the market, updated on a real-time basis. By driving forward the automation of its markets with the wide distribution of LIFFE CONNECT, its electronic trading platform, LIFFE offers customers a number of liquid futures products—primarily in the short-term interest rate and equity index categories.

Since a corporate restructuring in April 1999, the exchange has transformed itself from a membership organization into a commercial, for-profit, and customer-focused business. In conjunction with that change, LIFFE is creating separate but interdependent exchange and technology businesses that leverage off each other's skills and assets. The exchange business is committed to meeting customers' needs by providing trading opportunities on LIFFE CONNECT. The technology business is committed to providing electronic market solutions that bring together LIFFE's exchange, technology, and global distribution expertise. To help LIFFE realize its vision, it has secured a range of partners—including ISVs and network providers—to strengthen LIFFE CONNECT. To further realize these business objectives, the LIFFE Board selected three partners in June 2000, each of whom made a substantial investment in LIFFE's businesses. Cap Gemini Ernst & Young was chosen to help commercialize LIFFE's technology for a new business—a division of LIFFE (Holdings) plc—and be involved in the day-to-day development of the business and assist in its marketing strategy. Cap Gemini is one of the largest information technology consultancy firms in the world and is publicly listed on the Paris Bourse. The company offers management and IT consulting services, systems integration, and technology development, design, and outsourcing capabilities on a global scale. In addition, Battery Ventures and the Blackstone Group have invested significantly in LIFFE Holdings plc and provided extensive financial, technology management, and strategic development resources. Battery Ventures, based in the United States, is a global venture capital firm that creates and builds world-class, category-leading companies by assembling the key ingredients for success: management, technology, partnerships, and capital; The Blackstone Group is a New York based private investment bank, and since 1987 its merchant banking funds have raised over $6 billion for private equity investment.

The support of LIFFE's partners has provided the exchange with significant competitive strength that has led to an important partnership with the NASDAQ for single stock futures (SSF) traded on LIFFE CONNECT™. LIFFE, which did not have to deal with regulatory restrictions related to SSF, began trading them in 2000. LIFFE has begun to experience some success with the products and has great expectations for growth in the U.S. market when SSF begin trading there in late 2001. One of the interesting battles ahead in the fight for market share is likely to take place between the new LIFFE/NASDAQ entity and a similar type of entity formed by the CME, CBOE, and CBOT to trade SSF. While the stakes are high, it may not be a winner-take-all situation. It is quite possible that there will be liquidity in SSF on a number of exchanges.

Key LIFFE Futures Products and Clearing Fees

Short Sterling (£ 500,000)
Long Gilt (£ 50,000)
3-Month Euribor (£ 1,000,000)
3-Month Euroswiss (SFr 1,000,000)
FT-SE 100 Index (£ 10 per Index point)
Clearing Fee for all Financial Products (£ 0.50 per RoundTurn)
Clearing Fee for FT-SE 100 (£ 0.94 per RoundTurn)

THE eSPEED EXCHANGE (www.cx.cantor.com)

The eSpeed Exchange, in operation since September 1998, was the first fully electronic futures exchange approved by the Commodity Futures Trading Commission (CFTC), and the first new exchange approved by the CFTC in decades. The exchange is a joint venture between eSpeed and the New York Board of Trade (NYBOT). The eSpeed Exchange is a neutral trading platform and regulatory infrastructure that can be adapted for any marketplace. The trading platform can be used to create unique futures contracts that are designed by, and for, specialized marketplaces. The NYBOT is responsible for self regulatory oversight and clearing of all trades executed on the eSpeed Exchange. eSpeed operates the trading platform for the eSpeed Exchange, utilizing the trading and distribution systems that have been developed for the very liquid fixed income over-the-counter cash markets by the exchange's parent, Cantor Fitzgerald. The eSpeed Exchange's open-ended technology enables it to quickly launch new products as market demand dictates. In the future, the eSpeed Exchange will provide consortia, B2B markets, and individual companies with the ability to create their own, specialized futures and options contracts using eSpeed technology and the regulatory oversight capabilities of the eSpeed Exchange.

eSpeed's trading platform provides an algorithm for rules-based trading, with the goal of rewarding those who provide liquidity to the market. The Exchange promotes broad-based and direct access trading for all approved market participants. Entry barriers for direct participation on the eSpeed Exchange are minimal, allowing for more market participants and greater liquidity in the market. Direct participation is not restricted to members, but rather is open to anyone who registers with the Exchange and is guaranteed by a clearing member. This structure is designed to expand access and enhance liquidity. One feature that distinguishes the eSpeed Exchange is its innovative block trading rules. Block trading allows market participants to execute large trades directly with their customers and then post the trades at the Exchange to be cleared. This provides an important risk management tool for large market participants. Currently, the following futures products are traded on the eSpeed Exchange: U.S. Treasury Futures (2-year, 5-year, 10-year, and 30-year), U.S. Agency Futures (5-year, 10-year), and WI Futures (5-year and 10-year). The Exchange will be expanding its product offerings to provide futures and options products for a wide variety of marketplaces, including TradeSpark, a partnership of six major energy companies.

5
SOFTWARE

INDEPENDENT SOFTWARE VENDORS

The Independent Software Vendor, or ISV, creates the software that links the trader into the trading platforms of the exchanges through an Applied Programming Interface (API). The trader is able to see the prices that are trading and enter orders into the market through the ISV's software, which is also called the front-end. So, for instance, an ISV that "writes" to the API of the CME's GLOBEX trading platform allows the user to trade the e-minis on the ISV's front-end. There is a great deal of technology behind the front-end image on the trader's workstation, but for the vast majority of traders the ability to trade is appreciated in the context of the functionality (or lack thereof) of the front-end.

Many of the features traders like to use are available on all the major front-ends. Among these are one-click trading, one-click cancellation of orders, various market-making functions, configurability of the screen, real-time profit and loss information, and sophisticated risk management tools that allow a broker to monitor and control the customer's trading activity on a real-time basis. All of the major ISVs provide reasonably good functionality, but at this point there is no single front-end that is clearly the best. While there are subtle differences between them all—this one is better at one-click trading, that one has a superior spread trading function— a good trader can use any of the major ISVs to trade profitably.

There are two main distinguishing characteristics between the ISVs at this time. The first is connectivity to exchanges. A few ISVs are connected into virtually every major exchange in the world (e.g., GL has connections to over 40 exchanges worldwide). Others may only connect into the CME, CBOT, and EUREX. This does not necessarily mean one is superior to the other; after all, if one trades e-minis exclusively there is no particular reason to seek out a front-end that allows you to trade Polish equities. The second factor is price. There are wide disparities between the prices charged by the ISVs for their software. For the first few years of their existence, the ISVs typically charged a monthly license fee for the use of the software. For professional traders these fees could be thousands of dollars per month (depending on the number of exchanges to which one wished to be connected). Today, as monthly fees have fallen to between $300-600 for basic service, the ISVs have moved towards a transaction fee-based pricing model, charging a few cents per trade. At the present time, it is common to see a hybrid price, including a monthly license fee and a small transaction fee. It is likely that, very soon, the monthly license fee will disappear and the ISVs will rely solely on transactions to make their money.

Ironically, most traders will not directly choose their ISV. In most cases, end-users will be required to use the front-end that their broker supports. If one deals with the very largest FCMs it may be possible to have a wider choice, as a large FCM may support more than one ISV. If, however, you use a smaller brokerage firm, you are likely to be required to use that brokerage firm's ISV. Because there are so few substantive differences at this point between the functionality of the front-ends, it may be relatively unimportant which one the trader ends up using; after all, if one can trade profitably using any of the major ISVs, why spend too much time agonizing over this bell or that whistle? Still, there are two important considerations that must be addressed: first, your personal trading style may dictate the need for a particular feature, such as one-click trading or access to many exchanges. Therefore, you need to look for a broker that offers a front-end with those capabilities. Also, you must remember that the broker is going to pass along the cost of the front-end to the customer. This means that if the ISV is charging $500 per month and 10 cents per contract, you will pay that amount (or more) to the broker. Even if the broker decides to "waive" the license fee, don't be under any illusions; the waiver is simply a $500 soft-dollar rebate that the broker expects to make back in commissions. Complicating matters further, you will probably never know what the broker's deal with the ISV is, so you will not be able to know how much he is marking up the cost. The only thing that one can do to find the best deal is to shop around. As with other things for which one comparison shops, remember that the cheapest price is not always the best deal.

QUESTIONS TO ASK BEFORE CHOOSING A FRONT-END

Questions Traders Should Ask Before Choosing a Front-End

- Which exchanges can I access?
- Is the screen customizable? Can I change:
  - Cell sizes?
  - Font sizes?
  - Colors?
- How many clicks does it take to make a trade?
- Is there a single-click trading function?
- Can I cancel all outstanding orders with a single click?
- Can I view the entire "book" of orders residing on the exchange's trading platform or just a finite number of orders defined by the ISV?
- Is there a spread trading function?
- Can spreads be traded across exchanges?
- Can one trade using stops?
- Are positions tracked on a real-time basis?
- Can positions in multiple markets be aggregated?
- How many days of trading history can be viewed?
- Is real-time profit and loss information available?
- Can I obtain access to the front-end over the Internet?
- Is the Internet front-end a *thin client* (stripped down functionality)?
- Is there a monthly license fee for the software?
- Are there separate license fees for the direct connection and Internet connection?

- Does the ISV charge a per contract fee and if so, how much?
- Does the broker provide customer support or do I have to contact the ISV directly?

Questions Brokers Should Ask Before Choosing a Front-End

- Can I manage and monitor multiple users with multiple accounts trading multiple products at multiple exchanges regardless of my geographical location?
- Can more than one risk manager access the risk management piece of the software simultaneously?
- How sophisticated is the risk management functionality? Can I pre-determine:
  - Maximum position size?
  - Maximum trade size?
  - Maximum daily loss?
- Is real-time profit and loss information available?
- Is the information available in multiple currencies?
- Will risk management software alert me when a trader is approaching a pre-defined problem situation?
- Can I send messages to traders using the system?
- Can I disable a trader's front-end access?
- Can I link my proprietary software with the front-end through an Applied Programming Interface (API)?
- How does the ISV support the product?
- What hardware will I need to provide if I wish to create a direct connection to the exchanges?

THE VENDORS

This section provides a synopsis of each of the major ISV's offerings. Please be aware that information about the ISVs and their front-ends is constantly changing. By the time you read this, pricing as well as basic functionality could be markedly different. The good news is the cost will probably be lower, and the functionality is likely to improve.

Also, please understand that I am not endorsing one of these ISVs over the other. While I have my own preferences because of the active style of trading my customers engage in, I have included these vendors because I think that all of them are credible and can be used by good traders to trade profitably.

Trading Technologies (www.tradingtechnologies.com)

Trading Technologies (TT) was started in 1994 by Gary Kemp. Formerly an employee of Andersen Consulting, Kemp was one of the early architects of the DTB, later known as EUREX. Kemp, who serves as chairman and CEO, founded TT to provide a suite of exchange-independent trading products for the burgeoning electronic trading community. TT has since grown into a multinational sales and support organization of over 200 employees with offices in Frankfurt, London, New York, and Chicago. Recently, Andreas Preuss, a former EUREX board member, was appointed president of TT. Preuss is widely recognized for his role in the development of EUREX, the largest derivatives exchange in the world.

TT has become an industry leader in providing global exchange connectivity and high speed electronic trading systems. Often recognized for the speed of its software, TT has many adherents among very active traders. One of TT's users is Harris Brumfield, one of the most active electronic futures traders in the world. Brumfield, who for years was one of the biggest local traders in the pits at the CBOT, left the floor to pursue electronic trading when EUREX introduced its products to the U.S. market. Today, he regularly trades in excess of 100,000 contracts per day using TT and—like a modern-day Victor Kiam—was so impressed with the software that he invested in the company.

TT's product suite is best known for the high-performance front-end trading screen, X_TRADER®, which Brumfield helped to develop. X_TRADER was introduced for trading products listed on the world's leading derivatives exchanges (CBOT, CME, EUREX, LIFFE, and MATIF). It has recently been expanded to encompass trading in listed equities and other products. TT has outfitted X_TRADER with the intuitive market depth display known as MD_TRADER™ (see Exhibit 5-1). Users can now gauge market depth at a glance, in real-time, while trading directly from market activity. MD_TRADER offers a composite order book, position management, and easy order input/deletion capabilities. X_TRADER is a flexible trade execution tool, built for a wide array of users—from traders to institutional brokers.

Complementing X_TRADER in high-volume trading environments are X_RISK™ and X_QUOTE™. X_RISK™ enables risk managers to monitor trading activity of all traders and customers across

Exhibit 5-1 MD_Trader

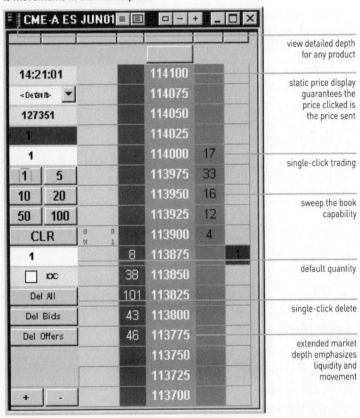

multiple products and exchanges from virtually any location. Key components to X_RISK™ include a full range of reports and alerts, flexibility to view positions by specific user-defined parameters, and the ability to view global trade positions in one easy-to-use database. With X_QUOTE™, traders have the capacity to generate a wide range of complex strategies and pricing schemes across multiple products and markets through one easy-to-use interface.

Trading Technologies offers products for a wide-scale distribution environment with X_TRADER® WEB™. Built from the market-proven technology of X_TRADER®, X_TRADER® WEB™ is a fast, secure, Internet-based trading system designed for massive scalability, bringing professional trading screen quality to the web browser.

Future Dynamics (www.fdlweb.com)

Future Dynamics (FD) is an United Kingdom-based Independent Software Vendor. FD's software applications include:

- Order Routing and Execution Systems
- Pre- and Post-Trade Risk Management
- Quotation Systems
- Online Analysis and Reporting
- Order Matching & Allocation
- Trade Reconciliation
- Internet-Based Management and Reporting

Crossfire, the FD front-end trading system, is a multi-exchange order routing, risk management, and execution tool built from Microsoft technologies and COM components. *Crossfire.Web* is FD's Internet trading application, and *PocketCrossfire* is a mobile PDA trading application, both allowing direct and real-time access to multiple markets while providing access to a shared global order book. The key features present in the Crossfire suite of products include:

- Full pre-trade risk management
- Spread matrix with implied and real-time functionality
- Outright trading from Spread Matrix

- One-click trading
- Quick trade—view depth of one's own trades and gaps in the market
- Shared Global order book
- Manual order input
- Charting and Technical Analysis
- Branding to comply with corporate image
- Scalable client-server architecture
- True Internet thin client allowing instant deployment
- Supports LIFFE, EUREX, MATIF, a/c/e, and the CME, with SGX and SFE in development
- Real time P&L—unrealized and realized
- Cash and margin based with clip and position size override
- Intuitive traffic light alert system
- Risk manager able to set cash warning limits

Exhibit 5-2 shows a screen shot of the *Crossfire* front-end.

Exhibit 5-2 Crossfire

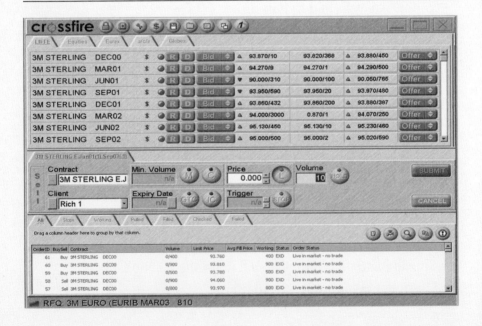

Pocket Crossfire

Pocket Crossfire is a multi-exchange order routing, risk management, and execution tool built from Microsoft technologies and COM components using wireless technology utilizing Internet standard dial-up and 3G telecommunications technology.

Key features include the following:

- Full pre-trade risk management
- Interactive order book with Crossfire server
- Real-time prices and order execution
- Configurable Exchange and Contract display
- One-touch trading
- Trader not tied to workstation throughout trading day
- Can be used to manage risk remotely

Exhibit 5-3 shows what the *Pocket Crossfire* application looks like, as seen on a *Compaq Ipaq* Personal Digital Assistant:

Exhibit 5-3 Pocket Crossfire

EasyScreen (www.easyscreen.com)

EasyScreen was founded in 1998 by two former traders from the London International Financial Futures and Options Exchange (LIFFE) following the exchange's decision to change its method of trading from Open Outcry to electronic trading. The EasyScreen system has been designed for traders by traders and offers sophisticated functionality while remaining easy, flexible, and intuitive to use. EasyScreen offers international coverage, having offices in London, Chicago, New York, Sydney, and Singapore.

EasyScreen's core product—EasyTrade—combines fully integrated functionality for futures, options, and equities trading on various electronic exchanges around the world, including:

- LIFFE
- EUREX
- Euronext Paris (MATIF)
- Chicago Board of Trade (a/c/e)
- Chicago Mercantile Exchange
- Singapore Exchange
- Sydney Futures Exchange

Future connectivity plans include the Cantor Exchange, Chicago Board Options Exchange, Euronext Paris (Monep), OM, and the London Stock Exchange (SETS). Connectivity is offered by a variety of methods, including via the Internet. The Internet products include both full "professional" and "light" versions of the EasyTrade front-end system, with the light model aimed at the retail market. EasyTrade features include:

- Access to multiple exchanges from a single screen
- Order types such as invisibles, icebergs, triggers, tranches, and baskets
- Sort orders by instrument, client, time, etc.
- Automated market-making
- A spread matrix offering a graphical way of tracking and trading Short Term Interest Rate calendar spreads, including implieds

- Screens and functionality fully customizable by the trader

EasyMinder, EasyScreen's pre-trade, real-time risk management system offers risk managers the ability to impose trading and financial limits on their traders and to monitor their performance in real time. The criteria against which a trader's performance is monitored may be custom designed by the risk manager, and can be tailored to suit the financial standing and the trading style of each individual trader.

EasyMinder key features include:

- Pre-trade order permissioning
- Permissions may be set in respect of: Position Limit/Profit and Loss/Net Equity
- Innovative "traffic-light" indicators to alert supervisors to problem situations
- "Window Explorer" style hierarchical tree facilitates trader, clients, trading groups, and sub-groups
- Pre-set permissions determine the trading and viewing rights of each node of the hierarchical tree
- Multi-exchange
- Complete audit trail log of activity
- Permissioning override facility
- Extensive clerical functions, including manual override for prices, edit cash balances, view, and edit trades
- Margin requirements may be factored up for higher risk clients
- Lock-out facility

EasyCell is a sophisticated trading tool, offering the facility to link the EasyTrade application to a customized *Microsoft Excel*™ spreadsheet and use the live exchange data as a basis for further calculation. Some of its many uses include the possibility to:

- Trade any hybrid strategy across different contracts and different exchanges
- Link to graphing tools to automatically trigger technical trading
- Automate Short-Term Interest Rate strip trades
- Work any trading algorithm that can be represented on Excel™

RealTime Systems (www.rtsgroup.net)

Formed nearly a decade ago, RealTime Systems (RTS) was conceived to efficiently route orders electronically to Ibis, the original electronic trading platform of the Frankfurt Stock exchange. Today, RTS connects to over 32 electronic exchanges and order routing systems worldwide and has a clientele of over 150 banks, investment firms, and trading arcades in Frankfurt, London, Chicago and New York, Paris, Amsterdam, and Sydney. Traders can, from a single workstation, trade multiple markets across numerous exchanges all over the world from a single interface.

RTS developed an open Application Programming Interface allowing its customers even more flexibility to customize their own trading applications and theoretical pricing models. Back office connectivity to the same API now offers RTS clients straight through processing capability for their trades. Continuing in its forward-thinking mode, RTS continues to innovate in every aspect of electronic trading, most recently incorporating the e-RTD Internet application into its arsenal of software.

GL Trade (www.gltrade.com)

GL Trade is the granddaddy of the ISVs. In operation since 1987, it currently has written to the APIs of more than 40 exchanges and boasts over 30,000 end-users. It has applications for professional traders in trading arcades as well as web applications for the non-professional trader. The key features in GL are:

- A Global centralization order system, called OPSYS, to allow global position keeping, administrative efficiency, and control.
- A market-making application, called GL Automate, for orders and Request for Quotes. Automate is able to automatically send bid and ask prices on the basis of an automatic analysis of the feeds integrated into the system and formulas determined and updated by the end-user, via an Excel spreadsheet.
- An application for active traders called Mosaic which allows for one-click trading.
- An Index arbitrage application called GL Arwin. Arwin automatically trades baskets of products determined by the end-user. A

simple trade function can generate orders on several instruments or markets.
- GL Spreadmatrix, which allows the end-user to trade up to 900 two-legged different instrument strategies from one window.
- Back office integration through an application called GL Glimp. GL systems are able to transmit real-time information to third party software allowing for links to middle and back office systems.
- Risk management through an application called GL Selector. Selector allows the risk manager to monitor in real time the activities of traders on the system, even in remote locations.

patsystems (www.patsystems.com)

Patsystems is a publicly owned firm based in the United Kingdom that has been creating trading software since 1994. Originally, the software was used to allow pit traders to trade on handheld computers. By 1995, the company was developing software to be used on the desktop off of the trading floor. The following features are found in the *pats* product:

- Access to multiple exchanges; at the time of this writing, *pats* connects to 13 exchanges globally, covering both derivatives and equity trading. In the future, *pats* will connect to additional exchanges as well as to the Over-the-Counter markets.
- Connectivity options: The broker has the flexibility to offer clients connectivity through the Internet or dedicated telephone lines.
- *pats* is a scalable system. This means that a broker can easily add any number of users to the system.
- Trade functionality: *pats* offers many high-end features in its front-end application. Among these are one-click trading, one-click cancels, the ability to trade cross-market spreads through a spread matrix application, single-screen access to multi-exchanges, and common order books. The software performs real-time profit and loss calculations, including closed trades and open equity, marked-to-the-market. P&L calculations can be performed in any currency. Positions and working orders can be tracked real-time. Five days of history can be displayed.

Pats has a highly developed risk management module in the software. All orders, whether they are routed via the Internet or through a direct connection, are initially submitted to a transaction server where they must pass through risk filters. There the market exposure is calculated on a real-time basis so that trades which are in excess of the limits pre-defined by the broker cannot be violated. Among the risk management features available are:

- Ability to manage and monitor risk over multiple users, multiple accounts, multiple exchanges and multiple products, regardless of geographical location.
- Comprehensive set of risk filters, such as maximum position limits, maximum traded size, maximum daily loss limit, and checks against initial margin.
- Risk managers have the ability to lock out traders, real-time. They can also send real-time messages to anyone on the trading network.
- Real-time profit and loss calculations in any currency. Access to customer profiles.

Interactive Brokers (www.interactivebrokers.com)

Interactive Brokers LLC (IB), formed in 1993, provides professionals and private investors with direct access to the world's financial markets. IB customers can directly route stock, options, and futures orders to 44 market centers in 13 countries. IB's worldwide access includes direct links to stock exchanges, including the NYSE, AMEX, and NASDAQ, as well as leading ECNs. All five U.S. options exchanges are also offered, as are international futures markets, including GLOBEX, a/c/e, Liffe, and HKFE.

Besides global connectivity, IB offers its customers flexibility in order routing. The customers can either route their order directly to the market center of their choice or elect to use IB's intelligent Best Execution technology. With Best Execution selected, IB's software continually scans competing markets and routes the order directly to the ECN or market center with the best prevailing price.

One benefit of the IB system is the availability of real-time streaming price quotes for any of the markets IB is connected to. When a stock, options, or futures symbol is entered on the system a real-time quote is displayed. The quotes are live and change continuously as the markets change—no periodic manual update or "refresh" is required.

Moreover, the system can display 24 separate trading pages, each listing 40 individual stocks, options, or futures. This allows the customer to monitor a total of up to 960 markets in real-time.

Other dynamic features of the Trader Workstation include:

- Account balance updates in real-time
- Quick, point, and click-trade execution
- Hot keys that rapidly route orders with a single keystroke

In addition to technologically advanced execution services, IB also offers clearance in several major markets at a wholesale price. IB only charges $0.01 per share, with a $1.00 minimum for stocks, $1.95 per options contract, and $2.95 per futures contract. This means IB customers can trade the all electronic e-mini products on the Chicago Mercantile Exchange's GLOBEX 2 system through IB for less than $6.00 a round-turn. All fees are already included in IB's commissions, and whether the customer places a limit or market order the price is the same. Additionally, market data fees for real-time streaming price quotes are included in the commission rates.

YesTrader (www.yestrader.com)

YesTrader (YT), formed in 2000, was started by three veterans of the CBOT trading pits. YT offers connectivity to eight exchanges, including EUREX, LIFFE, CBOT, and the CME. Its front-end application, called *Trans@ct,* allows the trader to trade or cancel all orders using a single click; see market depth going out to the 10 best bids and offers as well as other real-time market data; "get flat," or close out all positions, with a single click; and review profit and loss statistics in real-time. YT provides a risk management system called RiskWatcher (RW). The program allows the risk manager to:

- Pre-program trading limits and maximum position sizes
- View the real-time activity of any individual trader, trading group, or the entire customer base, irrespective of geographical location.
- Receive alerts in real-time based on pre-defined criteria programmed for each trader
- Warn or lock-out traders who are experiencing problems

YT offers two methods for accessing the software: the ASP solution or Client-Side solution. ASP stands for Application Service Provider and it works as follows: YT owns and maintains the servers upon which the software resides. The client hooks up his or her network by means of a secure network connection which allows him or her to use it. The ASP approach appeals to customers who do not want to assume the substantial up-front costs associated with buying, maintaining, and constantly upgrading network hardware. The Client-Side model is for those customers who wish to own and support their own hardware and connection. The customer provides the infrastructure and then YT technicians install it.

6
STOCK INDEX FUTURES

THE S&P 500 INDEX

Before we can begin to understand the characteristics of the S&P futures contract and the benefits of trading it, we need to explain what it represents and how it is used (For a complete list of the stocks in the Index, please see Appendix C). The futures contract is a *derivative* product, in that its value derives from an underlying index. The Index itself is *capitalization weighted,* which means that of the 500 component stocks, the ones with the largest relative worth have the most significant impact on the Index's movement. In other words, if IBM or General Motors goes up or down $1, the Index will fluctuate more than if the five hundredth largest stock in the Index moves the same amount. The money management community has embraced the S&P 500 as a benchmark because these stocks are large and liquid; have a market value in the neighborhood of 70 to 80 percent of the value of all the stocks on the New York Stock Exchange; and generally cover most of the important companies and industries in the United States. When stocks split, dividends are issued, companies experience mergers and acquisitions, or other material changes to the value of the component companies changes, Standard and Poor's adjusts the Index accordingly to maintain its integrity.

No money manager can ignore the pervasive influence of the Index on how the public perceives investment performance. If, for instance, the Index returns 25 percent in a given year, every fund manager in the country will be under pressure to provide a higher return. Many people, though, are confused about how to use the Index as a benchmark. In trying to determine whether a money manager has surpassed the Index, it is necessary to assign a risk value, known as *beta*, to the portfolio. If we say, for instance, that the beta of a portfolio is 1.5, it means for every $1 move in the Index, the portfolio will move $1.50. Similarly, if the beta is .5, the portfolio will move only half as much as the Index. Therefore, in order to make a meaningful comparison between a money manager's performance and that of the Index, one needs to consider not only the absolute return, but the *risk-adjusted return* as well.

THE S&P 500 FUTURES CONTRACT

The first thing you need to understand about the S&P 500 futures contract is that, unlike physically delivered commodities such as cattle, gold, or grain, the S&P is a cash-settled contract (For contract specifications please see Appendix B). This means "delivery" is actually a cash transfer of the difference between how much the contract sold for at the time of the original transaction and how much it is worth at liquidation. Let's consider how this works. First we need to determine the value of the futures contract. It is calculated by multiplying the futures price by $250. If, for example, the futures are trading at a price of 1000.00, then the value of the contract is $250,000. This means that if you buy a contract you effectively have control of that amount of stock for the duration of the trade (although you do not collect dividends). Each minimum trading increment, or *tick*, is .10 of an Index point which equates to $25 ($250 × .10). Therefore, if one bought the futures at 1000.00 and at delivery the price of the contract was 1050.00, the account balance would show a realized profit of $12,500 (50 Index points × $250). Of course, as with all futures products, it is not necessary to hold the contract until the delivery date. In fact, most trades are liquidated long before.

USING S&P FUTURES TO HEDGE PORTFOLIO RISK

No matter how carefully a portfolio is constructed, in a bear market it will almost always lose value. The existence of this *market risk* is the main reason that money managers use the futures contract: by hedging the value of their portfolios they can protect against losses in a market downturn. One might wonder, if the manager is convinced that the market is going down, why not simply liquidate the portfolio and reconstruct it when conditions become more favorable? There are a number of reasons why he might not want to do that, and they underscore the inherent benefits of using futures. If he sells the portfolio he may have to:

- Realize a loss or gain and suffer negative tax consequences
- Forego collecting dividends
- Generate heavy transaction costs, not only to exit the position but to reestablish it again when the market turns bullish

In addition, most money managers construct their portfolios with a long-term vision in mind. They may be willing to hold onto particular stocks in a declining market because they believe that they will eventually be rewarded for their patience. By selling futures against the portfolio, the manager gets the best of both worlds: he can keep his portfolio intact *and* insulate it from market risk.

Consider how this works. Let's say a manager has a $25 million portfolio he wishes to protect and the futures are trading at a price of 1000.00. In order to provide full coverage, the first thing he needs to know is the beta of the portfolio. Beta tells us how much we can expect a portfolio to fluctuate relative to the Index. We'll assume a beta of .5, which indicates that the portfolio is half as risky as the Index. Our *hedge ratio* becomes .5. Now we can figure out how many futures contracts the manager will need to hedge his exposure. We divide the value of the portfolio by the Index price multiplied by $250 (you'll recall that each Index point is valued at $250). Then we multiply the result by the hedge ratio and, voila, the manager knows the correct number of contracts to sell:

$$\$25,000,000/(1000.00 \times \$250) \times .5 = 50 \text{ contracts}$$

Similarly, exposure to an upside market move can be hedged. If, for example, a manager needs to buy stock in the future, he can protect his portfolio by purchasing futures. His calculation is a mirror image of the portfolio owner: the portfolio amount multiplied by the Index $\times$ $250, then multiplied by the beta, tells the buyer how many contracts he needs to purchase in order to construct the prospective portfolio free of market risk.

As you read these examples—which are quite simple—you may mistakenly assume that creating a hedge is a somewhat trivial pursuit.

You should understand that there are a number of risks involved in hedging and that not every manager chooses to pursue these strategies. The most significant risk is that the beta calculation is incorrect. Beta is not a constant. Because it derives from the past performance of the stocks in the portfolio it changes over time. If the manager calculates the beta incorrectly, he may execute too many or too few futures contracts to adequately protect the portfolio. Let's take a look at how a discrepancy in the beta calculation might affect our manager in the example above. In the original calculation we determined that the manager needed 50 futures contracts to hedge his position. If, however, the beta were really .75 he would need 75 contracts to hedge correctly. The practical result of the miscalculation is that for every point the futures contract falls the manager loses $6,250 ($250 × 25 contracts). The next potential problem is known as *basis risk*. While there is a close correlation between the futures and Index, the futures price move may be slightly different in percentage terms than that of the Index. Typically, basis risk will not result in the large losses that may be seen in cases where the beta calculation is wrong. Finally, there are some costs associated with putting on a hedge that must be accounted for. If, for instance, the market rallies after the sale of the futures, a manager will have a gain on the value of his stock portfolio and a loss on the value of the futures position. The profit and loss should cancel each other out, but the manager may have a different problem. Because futures are *marked-to-the-market,* the futures losses must be settled the next day in cash. The manager may be forced to borrow money to finance the position or liquidate his stocks, futures, or both. Such an outcome is clearly the antithesis of what the manager hopes to accomplish by establishing the hedge in the first place.

FAIR VALUE

Although you may never have traded a futures contract, if you follow the equities markets, and particularly if you watch CNBC, you will have heard the phrase *fair value*. The concept of *fair value* arises out of the symbiotic relationship between the futures and underlying cash Index. While buying or selling the futures contract serves as a proxy for buying or selling all 500 stocks, the price of the futures contract will invariably be different than that of the Index. The reason for this is that one who actually buys the component stocks receives dividends and must either pay for the stocks in full or take on interest expense for the margined total. The purchaser of a futures contract, by contrast, receives no dividends, but has to deposit far less than the stock purchaser in order to margin the position. The difference can then be invested in an instrument that provides a risk-free return, such as Treasury bills. Therefore, the main determinant of the gap between the price at which the futures and Index trade is the difference between the dividend return and what one can earn from Treasury bills or other similar investments. If dividend returns are low relative to credit market yields, then the Index will tend to trade at a large discount to the futures. As dividends increase and/or rates decrease, the premium of futures over the Index will tend to dissipate. As the futures contract moves towards expiration, the impact of the yield differential diminishes. Accordingly, the gap between the futures and Index will narrow and converge at expiration.

How to Calculate Fair Value[1]

The following formula is used to calculate fair value for stock index futures:

$$\text{Cash} [1 + r (2/365)] - \text{Dividends}$$

This example shows how to calculate fair value for S&P futures:

September S&P 500 futures price	1428.00 points
S&P 500 Cash Index	1414.00 points
Financing Costs/Interest Rate	6.0 %
Dividends to Expiration of futures (converted to S&P points)	3.32 points (see dividend calculation below)
Days to expiration of December futures	78 days
Fair Value of Futures	= Cash $[1 + r (2/365)]$ − Dividends
	= 1414.00 $[1 + .060 (78/365)]$ − 3.32
	= 1428.81
Amount of futures under-pricing	= 1428.81 − 1428.00 = .81

Dividend Yield Calculation:

S&P 500 Dividend Yield	= 1.10 %
Conversion to S&P points	= 1414 × .011
	= 15.6 points per year (78/365)
	= 3.32 points

---

[1] Chicago Mercantile Exchange, *Equity Index Futures & Options Information Guide* (Chicago Mercantile Exchange, Inc., 2001), p. 21.

Don't get intimidated by the math here. You will probably never need to calculate fair value yourself. Nonetheless, it is useful to understand how it is calculated. For most purposes, it is sufficient to listen for the number on CNBC before the market opening. If you miss that, you can find it, as well as a list of the fair value calculations for the previous six weeks, at the following web site: *www.allstocks.com/html/fair_value.html.*

INDEX ARBITRAGE (PROGRAM TRADING)

Theoretically, the yield differential that governs the relative value of the futures to the Index should dictate exactly how far apart the two instruments trade. In the real world, however, the spread between the futures and Index fluctuates, sometimes dramatically. When that occurs—and it happens quite often—a trading opportunity exists for sophisticated traders who are known as *arbitrageurs,* or *arbs.* The arbs look to capitalize on the disparity in prices by buying the under-priced instrument and/or selling the over-priced one. The reason this strategy works is because, as we've stated, the futures and Index prices converge at expiration.

It is this *index arbitrage,* sometimes called *program trading,* that enforces the fair value relationship. Consider how this works. If, for instance, a large speculator is bidding up the futures price while there is no similar interest in the underlying stocks, there would be a tendency for the gap between the futures and Index to grow in favor of the futures. As the futures to Index premium increases, the arbs look for a point at which they can safely sell the futures *above fair value* (meaning that after factoring in commission costs and execution risk they will still sell above fair value). They then buy the component stocks, knowing that with convergence at expiration they will be able to pocket the difference between the premium price at which they sold and the fair value price. Arbitrage enforces the fair value relationship because as the arbs pursue the transaction upon which the strategy is based—for instance, buying the undervalued stocks against futures—they *force the stocks higher and the futures lower,* thus bringing the relationship between the respective prices back to a level reflective of the fundamentals.

Index arbitrage in its purest form—executing transactions in all 500 stocks against the futures—is practiced by only a few large institutions. Only the most sophisticated and well-capitalized traders can afford the transaction costs and execution risks associated with such activities. There are, however, other market participants who perform similar types of transactions using stripped down baskets of stocks. The point is that there are almost as many different types of arbitrage strategies as there are arbitrageurs and no two arbs are exactly alike. The arb's strategies generate a regular stream of customer orders in the market which add greatly to the liquidity of the futures contract.

Most active futures traders pay close attention to fair value in order to try to figure out when program trades are likely to be executed. It would be nice if figuring this out was as simple as knowing that if fair value is x, then at a discount or premium of y the programs kick in, but that is hardly the case. Because there are so many different arbitrage strategies, it is impossible to know exactly when and how the arbs will come into the market. This doesn't stop traders from trying to discern when a program is imminent and the extent to which the market will react to the resultant buying or selling. One advantage that floor traders have is that, generally, they know which pit brokers represent the trading houses that execute programs. By closely watching and listening to these brokers—their body language, the timbre of their voices as they scream their bids and offers, and the urgency with which they try to execute their orders—the floor trader is able to acquire valuable information that can assist in the decision-making process. While the electronic trader does not have the ability to see the activity in the pit, it is definitely possible to achieve a high level of proficiency in ascertaining when the programs are being executed. One way to bridge the information gap between the pit and the screen is to listen to the actual prices trading in the S&P pit through a squawk box service. What you will hear is a member of the exchange who stands near the pit and reports the bids, offers, and trades as they occur (e.g., "20 bid on ten, ten offered at 40", "40 is trading"). Typically, she will also give market commentary: "20 bid on a hundred by Morgan Stanley, locals are lifting their offers; Morgan bids 40, locals are at a half; Half trading." By listening to the flow of information and carefully watching the prices trading on your screen, you can substantially replicate the pit experience, without having to be there. (Squawk box services are another example of why it is far more cost efficient to trade in an arcade. The service can cost between $300 to

$1,500 per month. For an individual the cost may be prohibitive. Spread out over many traders, however, it is nominal.)

Like snowflakes, no two programs are exactly alike. As a result the market reacts differently every time that a program is executed. Sometimes a program won't move the market a bit and other times programs will whip the market around as if it were a rag doll. Every successful stock index futures trader I know has developed an intuitive sense about fair value and program trading. The challenge is to develop a sixth sense of your own. It sounds like a daunting task—and in some regards it is—but if you know where fair value is and has recently been, watch the market trade on your screen every day, and listen carefully to the squawk box commentary, you are likely to become very proficient in understanding the ebb and flow of market forces.

THE TICK INDICATOR

One indicator that can help you get a sense of the market's broad movement is called the *tick*. Learning how to read and understand this indicator is particularly useful for those employing short-term trading strategies. The definition of the *tick* is the difference between the number of New York Stock Exchange stocks trading at a higher price than the previous trade and the number trading at a lower price than the previous trade. In other words, the *tick* is the number of up-tics minus down-tics. Generally, as the *tick* moves beyond +1,000 it indicates a bullish situation; as it moves beyond −1,000 it indicates a bearish situation. A scenario in which the *tick* approaches 1,500 (plus or minus) is a sign that the market is moving strongly in a specific direction. More important than the absolute number of the stocks higher or lower at any given moment is the relative value of the *tick* during the day or over a series of days. Whether, for instance, the *tick* is at 200 or −200 doesn't tell you much about whether the market is particularly strong or weak. If, however, the *tick* plunges very quickly from 200 to −200 this could be seen as an extremely bearish signal. Like all market indicators, the *tick* is subject to interpretation and every trader may construe its meaning differently. Nonetheless, it remains one of the best short-term trading indicators for those who wish to trade the S&P futures and e-mini contracts.

THE NASDAQ 100 INDEX

While the S&P 500 Index and Futures contract serve the needs of the traditional money manager, in recent years, increasing attention has been paid to the technology sector by both managers and speculators. For those market participants, the benchmark product is the NASDAQ 100 Index. The NASDAQ 100 futures contract, like the S&P 500, is traded at the CME (The complete list of stocks in the NASDAQ 100 Index appears in Appendix C).

The principle behind the NASDAQ 100 is the same as the S&P 500, but the characteristics of the product are very different. First, the NASDAQ is a smaller contract. To figure its value, multiply $50 × the NASDAQ Index price. Therefore, if the Index is trading at 1500.00, the contract is worth $150,000. Also, the minimum increment is .50 Index points or $50 per *tick*. The main difference, however, is that the NASDAQ 100 is far more volatile—2 to 3 times on average—than the S&P 500. This makes for an incredibly wild futures contract.

THE E-MINI S&P 500 AND E-MINI NASDAQ FUTURES CONTRACTS

Just as the CME pioneered the introduction of financial futures in 1972, so too did they initiate electronic futures trading in the United States. In 1987 the exchange announced that it was going to develop an electronic trading platform in conjunction with Reuters, and five years later the first contracts were traded on GLOBEX. Unlike financial futures, however, electronic futures trading grew slowly: the exchange—concerned about cannibalizing business from the pits—did not expand distribution rapidly enough, the technology was not very good at the outset, electronic products were only offered when the pits were closed and the markets were illiquid, and the cost to trade was too high. Any of these reasons by themselves, and certainly the combination of them, acted to constrain the growth of electronic trading. Even as EUREX began to experience explosive growth in a completely electronic environment, and the Sydney Futures Exchange, LIFFE, and MATIF closed down their trading floors, the CME and other U.S. exchanges continued to transact more than 95 percent of their business in the pits.

In 1997, however, the seeds were planted at the CME for the creation and growth of one of the most important and revolutionary exchange traded products in the world: the e-mini line of futures contracts. Interestingly, the e-minis almost never existed. At that time, the CME and CBOT entered into a bidding war against each other to try to entice the Dow Jones Company to license its Industrial Index as a futures product. It was thought by both exchanges that in a world where it had become increasingly difficult and expensive to introduce successful new trading products, the Dow was one of the few brand names which was worth paying a premium to own. Furthermore, in addition to whatever trading volumes and opportunities they believed would arise out of owning the Dow franchise, both exchanges felt that losing out to the other would cause a loss of prestige resulting in irreparable harm to the loser. As it became clear that the CME would lose the battle, it decided it had to respond to the devastating loss with a positive act. Accordingly, it implemented Plan B: introduction of the fully electronic e-mini S&P 500.

Almost from the moment it began trading, the e-mini started paying dividends. Not only had the CME avoided anteing up tens of millions of dollars to Dow Jones in license fees, but it had found a way—however inadvertent—to introduce electronic futures products that actually complemented the products in the pit. (Ironically, the CME had no intention of listing the Dow futures electronically. The CME architects even drew up plans for a large Dow pit which would be situated next to the S&P pit to facilitate spreading between the two instruments.) This element ensured the success of the new contract and laid the foundation for the explosive growth that would soon come. With the start of e-mini trading, the CME installed about 35 GLOBEX terminals surrounding the circumference of the S&P pit to facilitate member activity. This approach, while somewhat costly (although far less than the reported $50 million the CBOT paid to win the contest for the Dow), allowed traders who were curious about electronic trading or those who were simply fed up with the lack of physical space available in the S&P pit to try out the new market. Within a year after the product launch, members were fighting over who could have access to the terminals. The exchange gladly underwrote the expense of adding dozens more workstations for those who were waiting in queue.

There are a number of reasons that the e-minis have become so successful in such a short period of time. Chief among them, of course, is that they are electronic. Users enjoy the ease with which they are able to enter transactions and receive executions as compared to sending an order to be filled in the S&P pit. Being electronic, however, accounts for only part of why the e-minis have flourished.

Size Does Matter

The size of the contract—its *mininess*—makes it a far more tradeable product for the majority of users. You will recall that the full size S&P futures contract is very large. In fact, at the height of the bull market in 1999, the contract was worth in excess of $400,000. By itself, the sheer size of the contract does not create a problem; after all, a Eurodollar future is sized at $1 million per contract and it is the most heavily traded in the world. But the S&P is extremely volatile and the Securities and Exchange Commission, Congress, and the various stock and options exchanges have pressured the CME to keep margins on its Index products inordinately high. The exchange has fought this battle for years, and continues to seek relief, but thus far has been forced to capitulate on the matter to these powerful regulators and competitors. Unfortunately, most retail traders and even some professionals find the heavy margin burden a serious constraint on the ability to trade the contract. The e-minis solve this problem, because at one-fifth the size of the big contract they require only one-fifth the margin. To the CME's credit, it understood that big and wild was a bad combination, but small and wild provides an explosive bang for the buck. The mini size has been especially important in helping the e-mini NASDAQ contract grow. The full size, pit-traded, NASDAQ 100 contract, which is so much more volatile than the S&P that traders call it "the S&P on steroids," is far too expensive a contract for most users. Accordingly, while the volume in the e-mini S&P is usually 1.5 times that of the large size contract, the ratio of e-mini NASDAQ volume to its large contract is regularly 7:1. (In recent years, the exchanges have tried to distance themselves from the phrase "margin" in favor of the more descriptive "performance bond." Exchange leaders want the public to better understand the distinction between the margin that is put up to collateralize positions in the securities markets and due within three days after a transaction, and a performance bond in the futures market which must be available to maintain the position with the close of every trading session. For the exchanges, this is more than an exercise in semantics. They wish to impress that the futures markets are safe—safer perhaps than the securities markets—and pose no systemic risk. Few within or outside the futures industry, however, have weaned themselves from the traditional phrase. Consequently, it looks as if the word "margin" will remain in the forefront, while the phrase "performance bond" will stay on the margins.)

Location, Location, Location

The other significant factor that helped the e-mini products grow is the physical juxtaposition of trading workstations around the perimeter of the pit. When constructing the physical layout of the trading floor to accommodate the electronic traders, the exchange envisioned, correctly, that if sight lines to the pit were good it would be possible for the traders on the workstations to conduct an arbitrage between the order flow on the screen and that in the pit. Consider how this works: An electronic trader logs on at the start of the trading session. She carefully watches the activity in the pit but pays special attention to the broker wearing the navy blue trading coat of Goldman Sachs. About an hour after the opening bell, with the market trading at a price of 1250.00, the broker, who is known for moving the market when the Goldman trading desk in New York enters an index arbitrage order, suddenly puts her hands in the air. The electronic trader immediately tenses, knowing that something significant is about to happen. In fact, Goldman New York has just initiated an Index arbitrage trade and has instructed the pit broker to buy 200 futures contracts in the pit up to a maximum price of 1255.00. The electronic trader doesn't know how many contracts Goldman's order filler has been told to buy or how high to pay. What she does know, because she sees and hears it from her perch about 100 feet from the pit, is that the Goldman broker is frantically bidding up the market. The locals in the pit, who under normal circumstances make a tight bid/ask spread, also understand what is happening and lift their offers. In fact, smelling an opportunity, they join Goldman on the bid, which adds fuel to the rally. The electronic trader reacts by trying to buy the e-mini on the screen. Normally she would place a bid on the screen and hope that a seller came to her price—not this time. She is certain that the market is about to explode to the upside, so she is not going to waste time placing bids that are unlikely to be hit. Instead, she looks for an offer to reach out and buy. Because she is quick, she snaps up 30 contracts—equivalent to six full-size contracts—offered at a price of 1252.00. Meanwhile, back in the pit, the Goldman broker is still bidding up the market: 1252.00 . . . 1252.50 . . . 1253.00. The electronic trader smiles broadly because she is now riding a winning trade and about to bring it home. She will do one of two things: she can either execute a sale on the screen for 30 e-minis or she can flash an order to sell 6 S & Ps in the pit. In both cases she books a profit of 1 Index point on the position or \$1,500 (6 × 1 S&P Index point of \$250= \$1,500 or 30 × 1 e-mini Index point of \$50= \$1,500). This is only the beginning of the trader's fun. Never taking her eyes off of the Goldman pit broker, she looks for signs that the rally is running out of steam: Is the Goldman broker still actively bidding? Are locals in the pit still bidding? Are the other order fillers receiving new orders to buy from their customers? Are the traders starting to offer the market down, trying to take profits on contracts bought early in the rally? When the electronic trader senses that the buying frenzy has subsided, she knows it is time to act. She doesn't waste time offering—once again, this is a situation in which she can't take the chance that she will miss the move—and quickly searches for a bid to hit on the screen. The market is falling fast: 1254.00 . . . 1253.50 . . . 1253.00. She enters an order and is able to sell 30 contracts at 1253.00 and is now short the market. Suddenly it seems as if everyone in the world wants to sell. The price falls farther: 1252.50 . . . 1252.00. The electronic trader can't stifle her smile as she prepares to collect her winnings yet again. She flashes an order into the pit to buy 6 S&Ps at 1252.00, and when the order is executed she collects another profit of \$1,500 . . . not too shabby for 90 seconds of work.

While this example is an oversimplification, it is a quite realistic portrayal of the way a series of e-mini trades might take place on the floor of the CME. Please, however, do not get the wrong impression about the skill required to carry out such transactions. Electronic and pit traders must compete for every transaction against dozens or hundreds of other traders. Only the quickest, most prescient traders are able to make winning trades, like the ones previously explained, with any degree of regularity.

Throughout the trading session, this arbitrage drives the volume in the e-mini and S&P contracts. Does the pit lead the e-mini or vice versa? When the e-mini was first introduced it was clearly the pit that dictated the action. Now, however, the link between the two contracts has evolved to the degree that no one can tell anymore which contract is the chicken and which the egg. While there is probably a tremendous trading strategy waiting to be uncovered with the correct answer to the question, for our purposes it is sufficient to note that the symbiotic relationship of the S&P and e-mini creates an ample supply of liquidity in both these products.

SO YOU WANT TO BE A MARKET-MAKER

It is no wonder that so many traders would like to emulate the market-makers. Market-makers tend to be highly profitable and extremely consistent; they seem to buy slightly before demand heats up and are out the door just a bit before the crowd rushes for the exits. Every trader I know is trying to achieve that level of trading dexterity. Market-makers are actually employing a simple principle: By studying and reacting to the *momentum* of the market, they can capitalize on the many wiggles and jiggles that make up a typical day's price movement. Market-makers do not particularly care whether the market is heading up or down; they can make money in either environment. In up markets they look to buy on breaks; in down markets, they sell on rallies. In a way, this seems counter-intuitive; after all, everyone knows that the "trend is your friend." If a market is breaking, wouldn't you want to sell it? Similarly, if a market is rallying it seems as if selling it is that last thing you would want to do. The logic behind the approach is based on the premise that in the short-run a market will not continue to trade in a particular direction without stopping and retreating slightly; in a market like the e-mini S&P or e-mini NASDAQ there are likely to be many points in every trading day at which the market is saturated with either buyers or sellers. It is these opportunities that market-makers look for. They can fade the immediate move because they realize that at a saturation point there is relatively little risk that the market will move against them. In fact, there is a high probability that the market will move in their direction and everyone else who traded at or near the saturation point will have to go to the market-maker to get out of the losing position.

This trading approach is familiar to anyone who has traded NASDAQ stocks with a Level II system. There, as in the futures markets, market-makers constantly place bids and offers on the market and wait for other market participants to trade with them. When they sense that a market is turning, the market-makers do everything they can to try to get on the right side of the move and are usually able to do so. Let's say that the e-mini S & P contract has been rallying all day and is up over 1,000 points from the previous day's settlement and 500 points above the previous day's high (very bullish signals). The market-maker is primarily going to be looking to buy the market on a day like that. He will not, however, necessarily buy every offer he sees. He is more likely to buy the market every time it experiences a 200 point pullback. In fact, he may even try to fake out the other market participants in order to accomplish his goal. As the market breaks back to the point at which he'd like to establish a long position, he may place a large sell order simultaneously with his bid. He knows that some of the weaker traders will see that offer and think the market is going into a free-fall. Traders who are long may be fooled by this trick and flatten their positions, perhaps even reversing and going short. I can tell you, as one who has been caught on the wrong side of this trade many times, there is nothing more frustrating than knowing that the market-makers took your money (yet again). You feel stupid, worthless, and mad. The only way to get back at them is to beat them at their own game; like Bill Murray's character said in *Caddyshack,* as he contemplated how to get even with the gopher who was burrowing underneath the Bushwood Country Club golf course, "You've got to think like a varmint, you've got to become a varmint."

TRADING THE "LEVELS"

A level can be defined as *the price area that includes both the market-maker's bid and offer.* As we've said, markets seldom move in a specific direction without making at least a small pullback or rally. The key to the market-maker game is in understanding how many levels, on average, the market will move before the pullback or rally. To the extent that you acquire a feel for this, you will begin to think like a market-maker. You'll become a market-maker. Bill Murray would be proud.

Let's consider another example of how a market-maker plays the levels. By watching the market closely over a period of days during which the S&P contract is trending sharply lower, it seems that the downside momentum pauses with every 1,000 point break ($500 per e-mini contract). With each pause, bargain hunters enter the market and buy the offers, causing a short-term rally of 500 points. The astute market-maker may buy along with the bargain hunter, but what the market-maker is really interested in doing is determining the saturation point of the rally and selling as many contracts as possible when it is reached. She knows that when the inevitable break comes, she will be short at the top of the market. Getting out with a profit will be easy because the longs, who stupidly bought at the saturation point, will beg the market-maker to show a bid.

You are probably wondering how in the world you're supposed to accomplish this. I wish there were some formula I could impart, but there isn't (actually, if such a formula existed and I had the secret, chances are I'd keep it to myself). The only way I know to learn how to read the levels is to study the market diligently—watching the screen and reviewing the charts until your eyes glaze over. This may strike you as boring or even pointless, but through the repetitive process of observing the market for endless hours, the truly perceptive among you will begin to appreciate the bands within which the market moves and then snaps back. Many will fail, either because of a lack of effort or ability. Some will develop the occasional intuition. But if you are among the few who are able to visualize which way the pendulum is swinging and where in its arc it is at all times, then you won't be needing any more books on how to trade.

THE PRIVILEGES OF MEMBERSHIP

Anyone who has ever belonged to a country club can appreciate that the proprietary assets of the group are used for the benefit of the club's members. Whether it is a choice of tee times, a reserved dressing space in the locker room, or access to any of the club's facilities, the needs of a member will always take precedence over those of a non-member.

This comes at a cost, of course, but the member is happy to pay dues and fees in order to enjoy the privileges of membership. It is the responsibility of the country club's management to spend its revenues wisely to support the facilities; to create and maintain a comfortable environment, including, for example, a well-manicured golf course; a pool that's heated to just the right temperature; and a dining room that turns out a Sunday brunch to die for. Management's job, in essence, is to promote the "liquidity" of the country club.

The CME operates very much like a country club when it comes to e-mini privileges. While anyone can trade the contract from his or her own PC, before one can trade on the floor overlooking the pits he or she must buy or lease a membership at the exchange. This ownership confers certain rights that have contributed to the success of many members. For instance, the exchange provides trading workstations on the floor at little or no cost. Similarly, there is no charge for using the high-speed trade matching network created by the exchange. Members who are extremely active enjoy the biggest benefit of all: a $50 per day cap on the GLOBEX fees they pay to the exchange and substantially reduced clearing fees (clearing fees are separate from GLOBEX fees). For an individual who trades 1,000 contracts per day (not an unreasonable amount), the cap results in a $450 per day savings in GLOBEX fees, or more than $100,000 per year. In clearing fees the savings is even greater, as seat owners pay $0.63 less per round-turn transaction than non-seat owners. In fact, the decision to get a "seat" at the exchange is often based on the answer to this question: Do the fee breaks exceed either the financing or opportunity cost of owning or leasing the seat? If the answer to the question is yes, obtaining the seat becomes a prudent business decision. (Both owners and lessees qualify for the $50/day cap, although only owners receive the clearing fee break. At this writing, seats lease for approximately $4,000/month.) To the extent that a trader is very active and interest rates are relatively low, the returns on the cost of capital to own the seat can be phenomenal. In periods when volume is heavy and where demand for seats is high, there is also the possibility of capital appreciation; if the seats are in short supply because active traders need to own them to achieve fee breaks, then the price will rise. Of course, what can go up can also come down. Seat prices tend to collapse when volume is low and trading opportunities diminish.

Clearly, the cost to the CME of providing these benefits to members is substantial. Some critics argue that it is really the customer who bears the cost of these benefits. They say that if the exchange forced members to bear a greater percentage of the cost, then it could afford to charge customers lower fees. The exchange, however, justifies the expense because facilitating the members' ability to trade actively is the foundation of the market's deep pool of liquidity. Consider this: Were the exchange to lift the member fee cap, would the 1,000-contract-per-day trader still provide the same amount of liquidity? Would he be constrained, over time, from doubling or quadrupling his daily volume? How much better off would the customer be with lower fees, but markedly reduced liquidity? These questions and a hundred others relating to the cost of providing liquidity are hotly debated at exchanges everywhere. They are certainly interesting—and will make good case studies for the next generation of University of Chicago MBAs—but they are germane to our interest only to the extent that we understand the following important principle: *All exchanges promote liquidity by underwriting the cost of trade facilitation for their members.*

E-MINIS: THE DAY TRADER'S DREAM COME TRUE

In looking at the other reasons that the e-minis have grown so quickly, it is important to realize that it is not only money managers, large institutions, and members on the floor of the CME that find these products so appealing. The inherent benefits of the e-minis make the products attractive to equities day traders as well. While most professional day traders have yet to emigrate from trading equities online, as the e-mini gospel is spread there is likely to be a substantial exodus from equities to futures. Let's take a look at some of the reasons.

Commission Savings

While commission rates for equities trades have come down considerably with the growth of online trading, the active day trader still assumes a fairly high commission burden. The best rates available are in the neighborhood of a penny per share. Therefore, a 1,000 share transaction—typically the default amount for most active traders—costs $20 for a round-turn transaction ($10 to initiate the transaction and $10 to close it). This may not seem like much, but consider that an active day trader may make 100–500 transactions per day. At those rates, the trader will pay his broker-dealer firm $2,000 to $5,000 per day. Futures, by contrast, are far less expensive. An active e-mini trader is likely to pay under $5 per round-turn.

Increased Leverage

Under Federal Reserve Regulation T, a stock trader's *buying power* using margin is limited to twice the actual cash or security value in her account. Let's say a day trader wanted to buy 1,000 shares of a $50 stock. In order to comply with "Reg T," she will need to have at least $25,000 in account equity in order to purchase the $50,000 worth of stock. By contrast, the margin on an e-mini S & P contract is $4,688, granting the trader far more leverage. With the e-mini at 1000.00 ($50,000 in value) and a $25,000 account the trader can control five e-mini contracts, or $250,000 worth of stock. Traders love leverage; the more they have, the lower the burden upon them to maintain a large amount of capital in their account. (In the first few years of the equities day trading boom, many broker-dealers tried to skirt Reg T by allowing customers to lend the unused buying power in their accounts to other customers who needed additional capital to trade. The SEC has come down hard on these broker-dealers and prohibited the practice.) The biggest benefit, however, is that successful e-mini traders use the additional leverage to make more money, more quickly, than would be possible with the lower leverage available in the securities markets. Of course, with additional leverage, unsuccessful e-mini traders tend to lose more money, more quickly too.

Easier to Use Technology

NASDAQ trading systems are intimidating. Because there are so many ways to transact business (e.g., SOES, SELECTNET, NASDAQ, ECNs), the trader must learn a long list of commands before he can trade. Moreover, he must learn to differentiate between the various types of orders and determine which is best suited for the immediate purpose. Many NASDAQ day traders suffer significant losses, particularly at the outset, simply because this learning curve is steep. Many traders learn the techniques they need to become successful, but others never recover from these early losses. Because there is a Central Limit Order Book on the GLOBEX electronic platform, to which all orders are sent and matched and from which all information is conveyed back to the trader, commands to trade the e-minis, by contrast, are far simpler to learn and use. Also, as discussed earlier, the CLOB makes futures trading far more efficient than trading securities on a NASDAQ Level II system.

Level Playing Field

Even as NASDAQ opened its market to the "non-professionals," as mandated by the order-handling rule changes in 1997, it maintained certain barriers that are inherently unfair to day traders. Among these are the following: a market-maker can decline to accept a day trader's order, putting the day trader at a severe disadvantage; Small Order Execution (SOES) rules were changed (after a strong lobbying effort by the market-making community), with the result being that day traders find it virtually impossible to satisfy their demand for stock through this type of a trade; and market-makers are permitted to pay brokerage firms for order flow which gives them information and profit opportunities not available to the day trader. When it comes to the e-minis, by stark contrast, every market participant is treated as an equal and every bid and offer on GLOBEX is available on a first-come, first-serve basis. What this means is that the ability to succeed depends not on who you are, but how good a trader you are.

No Short-Selling Prohibitions

The futures trader can sell a bid at any time to get short, unlike NASDAQ day traders, who must wait for an up-*tick*, sell on the offer, or commit substantial amounts of capital to establish synthetic short positions using options. The ability to get short at any time with a minimal amount of slippage is of tremendous benefit to the e-mini trader, particularly in times of panic selling. Also, in contrast to NASDAQ, which prohibits short-selling in many of its stocks (there is a daily list of stocks which cannot be shorted), the e-mini can be shorted at any time.

Large Daily Ranges and No "Decimalization"

The average dollar value between the high and low in the e-mini S&P contract is in the neighborhood of $1,500; the range in the e-mini NASDAQ product is about $3,000 (that equates to a $7,500 range in the big S&P contract and $15,000 in the big NASDAQ contract). Furthermore, the minimum price increment in these products is sufficiently large to allow for traders to make wide spreads between their bids and offers. This is in stark contrast to stocks traded on the NYSE and NASDAQ, where it is possible to bring the bid/ask spread in as tight as a penny.

Preferential Tax Treatment and Filing Requirements

Trading profits in the securities markets are considered short-term gains and are taxed at the trader's ordinary income tax rate. By contrast, gains on futures trading are considered 60 percent long-term and 40 percent short-term. The resulting blended rate of less than 30 percent means that a successful futures trader will pay substantially less tax on gains than would a successful stock trader. Additionally, the filing requirements for futures trading are far less cumbersome than for stock trading. Every stock transaction, even if there are thousands of them, needs to be accounted for. The IRS has no such requirement regarding futures trades.

TAX CONSIDERATIONS[2]

The preferential tax treatment afforded to futures traders is a significant inducement and potential benefit (although one should not lose sight of the fact that in order to appreciate the benefit you need to make a profit). For securities traders who are profitable and considering making a switch to futures trading, this advantage may be the clincher. For them, making the switch is similar to deciding whether to move from a state in which there is an income tax to a state which has none; if one

---

[2] While every attempt has been made to present accurate tax information, please consult a professional accountant or tax attorney about this matter, as the laws are subject to various interpretations and periodic amendments.

can make a living in either place (and all other things are equal), it is only prudent to move to the place that allows you to keep more of your money.

For those who are not professional futures traders, or are contemplating trading full-time, there are a number of issues that need to be understood in order to know whether you can benefit from this preferential tax treatment. You may wish to know the following: If I have another job, can I receive the preferential tax treatment? While there is no explicit statement in the law that one must trade full-time in order to qualify for the 60/40 treatment and, at this writing, no definitive case law that is widely applicable, if one is audited it will be necessary to show that in the time spent trading one acted as a professional trader acts. Among the questions that the IRS may ask are[3]:

- How many transactions do you average per day, month, and/or year?
- What instruments do you trade?
- How long have you been trading?
- How many hours per day do you spend on the markets?
- What percentage of your work day is spent on each of the respective jobs?
- Do you trade from a professional "trader's" office, even if it is in your home?
- Do you have distinct strategies based on various professional market approaches?
- Do you use a trading system?
- Do you maintain a library of trading books?
- How do you monitor your performance?
- How much do you spend on trading related items?
- What portion of your net worth does your trading account comprise?
- What is your income from non-trading related work?
- What is your trading income over the past year or two?

---

[3] Ted Tesser, *Active Trader Magazine* (November 2000), p. 96.

Although the case law gives us no definitive answer as to the "correct" number of trades to be considered a professional, it has told us how many trades are too few. A number of years ago, a gentleman named Rudolph Steffler argued that he was a full-time trader entitled to the 60/40 split. Steffler had no other job, but traded infrequently; 27 trades in three years. The court ruled that even though trading was his primary source of income, he had not traded actively enough during the period in question to be considered a professional. So it is probably fair to say that if you hope to be accorded the preferential tax treatment of a futures trader, it would be prudent to make more than nine trades per year. For a complete look at the tax regulations, see Appendix D, where the text of the rules is presented in its entirety.

SINGLE STOCK FUTURES

For at least 15 years, U.S. futures exchanges have been lobbying regulators to allow the trading of futures on individual stocks and, until recently, their pleas were rebuffed. Late last year, however, the perseverance of the exchanges was rewarded with the implementation of the Commodity Futures Modernization Act of 2000 (CFMA), a 262-page document that only a bureaucrat or attorney who charges $400/hour can appreciate. The CFMA will allow exchanges to list futures contracts on individual stocks as early as August of 2001 for professional market-makers and by December 21 for retail participants. While there is great debate as to whether *Single Stock Futures* (SSF), as they are called, will be embraced by the trading community, every single major exchange is gearing up for the battle to capture market share, if it develops. Among futures exchanges, the CME, CBOT, and NYMEX should lead the way, although it is not a foregone conclusion that the business will migrate to a futures exchange. Securities exchanges like the American Stock Exchange, New York Stock Exchange, and NASDAQ are extremely interested in competing as well and will not willingly cede the business to the futures exchanges. In fact, it is expected that futures and securities exchanges will join together in strategic alliances. At the time of this writing, the NASDAQ has announced a strategic partnership with the LIFFE to trade single stock futures, and the CME, CBOE, and CBOT have announced a joint venture. Other players, such as ECNs and brokerage firms, are likely to emerge as well, and it may take some time to figure out who is doing what and how and why they are doing it. One thing is certain: Many millions of dollars will be spent to try to become the leader in what some futures industry insiders believe to be the next great growth market in derivatives.

The idea of single stock futures is fundamentally no different than that of futures on any other financial instrument. The reasons that the regulators prohibited the exchanges from offering these products until now has to do with various competitive business issues between futures and securities exchanges and the internecine political struggles between the CFTC and SEC. Because the regulators could not agree as to who would govern the products and those who trade them, they simply prohibited the activity. To remedy this problem they finally agreed to share governance and determine together how to impose margins, tax gains, adjudicate disputes, and decide a litany of other sticky issues. In order to get the CFMA bill passed, a number of very tough issues were left open with the understanding that if the CFTC and SEC could not resolve them within a reasonable period of time that the right to trade SSF could be rescinded. It remains to be seen whether two governmental agencies with such heterogeneous interests can work together and effectively regulate what may become a very large and complex business.

Why Are SSF Needed?

There are significant benefits to SSF. Among them are:
- Providing investors with an easy way to move in and out of particular stocks without having to add or subtract those stocks from their cash portfolios
- Preferential margin treatment, allowing for capital efficiencies and increased leverage
- Cash settlement
- Creation of arbitrage opportunities for traders who wish to capitalize on the relative movement of the underlying securities and the SSF
- Enabling various combination trades such as one stock against another or a stock against a stock index, which may be difficult to consummate in the underlying cash market

- The ability to go short without having to borrow stock and without having to wait for an up-*tick*

Notwithstanding the CFMA's recognition that trading in SSF should be allowed and the substantial potential benefits to the trading community, there are still many hurdles to overcome before SSF can be considered viable. Aside from the fact that the bill delayed actual trading of the instruments for almost a year, it left open a number of potentially troublesome issues that even has industry experts bewildered. Among other things, it remains unclear how SSF will be margined, who can trade them, whether foreign exchanges can offer them to U.S. market participants, and whether users will appreciate the preferential tax treatment afforded to professional futures traders. From a business standpoint there are questions as well: Will there be liquidity? Who will provide it? Will a single exchange or multiple exchanges capture the market? Will non-exchanges, like ECNs, trading houses, brokerages, banks, and technology providers compete for a share of the market? Will brokers sell the instruments to their customers in lieu of securities? It is far too early in the process to even venture educated guesses to these questions given how convoluted the regulations are. Ironically, Section 2 of the CFMA states that among the purposes of the legislation are the following:

- Streamlining and eliminating unnecessary regulation.
- Providing a framework for allowing the trading of futures on securities, and promoting innovation for futures and derivatives by enhancing legal certainty and enhancing the competitive position of U.S. financial institutions. Unfortunately, as the regulation stands, it is hard to see how any of the stated goals can be met.

It will pay to keep an eye on SSF as the exchanges and regulators fight their battles, because eventually there may be great trading opportunities with the products. In the interim, however, it seems that the needs of most traders can be sufficiently met by trading stock index futures. If you are intent on being a pioneer, you will probably find the initial markets in SSF to be illiquid and volatile. Some traders are able to capitalize on the inefficiencies of markets such as these, but for most of us, this type of trading is akin to walking a tightrope without a safety net. My professional advice is to step aside at the outset and let others build the market's foundation.

7
FOREIGN EXCHANGE

FOREIGN EXCHANGE AND ELECTRONIC TRADING SYSTEMS

As noted earlier, when financial futures contracts began trading at the CME in 1972, the first products traded were currency contracts. At the outset, the major players in the foreign exchange markets—also known as FOREX or FX—tried to ignore the traders on the floor of the exchange, but with the abandonment of the Bretton Woods agreement that fixed exchange rates, interest in currency trading grew quickly among several groups: those with currency risk who used the exchange-traded products to hedge; arbitrageurs who traded futures against cash currencies; and finally, speculators who were attracted by the profit opportunities that came with the volatility in the marketplace. Before too long, the banks realized that there was a fortune to be made by expanding their foreign exchange operations and they became full-fledged participants in the FX trading revolution. They purchased exchange memberships, put brokers on the trading floors, and actively solicited retail business, particularly among commodity trading advisors and hedge fund managers.

Since the first financial futures were FOREX contracts, perhaps it is fitting that currencies were the first products to be traded electronically. The experiment began in 1992 when Reuters developed an electronic system that allowed market participants to communicate bids, offers, and messages over a network. Traditionally, the FOREX markets were traded over the phone and in the currency pits at the CME, but the Reuters platform was eagerly embraced. FX business began to migrate away from the accepted methods almost immediately and the liquidity of the Reuters Dealing System, as it was called, was hard for even the most recalcitrant FX traders to ignore.

In response to Reuters' success, a group of the world's largest banks—who had never been happy about having to share the FOREX pie with the CME or anyone else—saw an opportunity in electronic trading to recapture what they felt was its rightful preeminence in the FX arena. In 1993, they formed a consortium called Electronic Broking Services, or EBS Partners, and established an online network of FX market-makers. (The partners in EBS include ABN-AMRO Bank; Bank of America; Barclays Capital; Chase Manhattan Bank; Citibank; Commerzbank; Credit Suisse First Boston; Hong Kong Shanghai Bank; J.P. Morgan; Lehman Brothers; Royal Bank of Scotland; S-E Banken; UBS AG; and Minex Corporation.) They were very successful in capturing market share. Today, while Reuters maintains a strong presence in FX, particularly in the British Pound, Canadian Dollar, and Australian Dollar, EBS is the clear leader in Japanese Yen, Eurocurrency, and all the key currency pairs, with 2,500 users at 850 banks around the world. It is estimated that more than 85 percent of all FOREX business is done on EBS or Reuters.

The key thing to understand about EBS and Reuters is that these systems are only for banks and investment dealers. Customers, such as corporations and money management firms, are not allowed onto the systems. Typically, these entities will have a relationship with a bank that has access to Reuters or EBS. Then, when an FX price is requested, the bank will make a market for the customer based on the electronic price in this private market. The bank will almost certainly mark up the price to the customer to cover the cost of doing business and make a profit.

The introduction of Reuters and EBS to the market has had an interesting—and unintended—effect on the banks who participate on the systems. While they have effectively achieved their goal of eliminating non-electronic competitors, they have also created an extremely transparent market. To be sure, that customers do not have direct access to EBS pricing gives the banks a measure of control over the marketplace, but because pricing information is so centralized and flows freely, the banks are forced to compete with each other on pricing to the point at which there is only the tiniest bit of profit to be made on the bid/ask spread. As a result, many entities that made big money in the 1970s and 1980s trading FX have merged with the major players or gotten out of the business entirely. To listen to the surviving bank traders talk, one would think that they carry on their FOREX operations on a pro-bono basis. Do not believe a word of it. There are a lot of endeavors to which a bank can commit its capital, and if FX were not a profitable venture they'd leave it to the exchanges.

So where does that leave the rest of us who, like Milton Friedman over 30 years ago, may want to place a bet on the direction of a particular currency. In fact, there are more options than ever for the individual trader, and particularly for the electronic trader who wants to trade FX. However, we first need to understand more about how the cash FX markets operate and their role vis-à-vis the futures markets.

THE INTERBANK MARKET

The term *Interbank* is used generically to identify the foreign exchange markets. The participants in the *Interbank* market transact business in every major financial center in the world. The volume of trade is enormous; more than a trillion dollars a day by the most recent figures from the Bank of International Settlements (an organization that monitors the activity of currency market participants). The *Interbank* is almost always open somewhere in the world, and consequently its participants stand ready to capitalize on trading opportunities whenever and wherever they present themselves. While some business is conducted through the auspices of foreign exchange brokers who broker transactions between end-users and banks, the vast majority of the business is transacted on the Reuters Dealing System or EBS. Exchange-traded currencies, such as those traded on the floor of the CME or the Dollar Index at the New York Board of Trade, are not considered part of the *Interbank* market.

THE FOREIGN EXCHANGE DEALER

While some banks in the *Interbank* market may act as agents, charging their customers a commission for each transaction, most also act as principal to the trade, offering the customer a two-sided market with enough of a spread built in to provide a profit margin. The dealer must work hard to cultivate his customer base by providing a tight bid/ask spread and deep pool of liquidity. These are the various ways in which the dealer might make a trade:

- Bank X contacts the dealer and asks for a price on $10 million worth of Canadian Dollars. The dealer checks the prevailing bid on Reuters and, not knowing for certain whether Bank X is buying or selling the currency, quotes a price of 1.5015/1.5020. The dealer is confident that there is enough of a spread between the bid and ask so that whether Bank X buys or sells it will be possible to make a profit. Bank X, liking the dealer's price, sends an electronic message agreeing to sell the $10 million at the bid price of 1.5015. In all likelihood, the dealer has already lined up another dealer on the system who will make a market on an equivalent amount of Canadian Dollars that allows him to close out the trade with a profit.

- A large hedge fund manager, who is a customer of the dealer, puts in an order to buy $100 million worth of Japanese Yen. The dealer will search for an offer on EBS and execute the order. He may choose to take the trade into his own portfolio, although he is not necessarily required to do so.

- The bank that employs the dealer initiates a transaction to satisfy its own internal operations, asking the dealer to sell 20 million pounds Sterling at 1.4850.

- The European Central Bank unexpectedly raises interest rates, making the Eurocurrency more valuable relative to the dollar. The dealer, thinking the market will move sharply higher, buys $25 million worth of Eurocurrency for the proprietary account of the bank. If the dealer has a resting customer order to sell Eurocurrency, she is likely to execute the customer's order by buying it for her own portfolio. (In the futures markets, which are regulated by the CFTC, trading against a customer order is expressly prohibited. The CFTC argues that a broker who acts as both agent and principal has an inherent conflict of interest. In the *Interbank* market, which is unregulated, there are no such prohibitions.)

- In an active, volatile market, the dealer generates revenue for the bank's proprietary account by scalping tiny profits off of the bids and offers of other dealers on the system. To the extent that the dealer has sufficient credit to deal with many counterparties and is facile enough to beat the other dealers at their own game, this form of trading can be very profitable.

While FX dealers and traders have a somewhat shady reputation in the financial world—central bankers and countries with weak currencies in particular revile them—the *Interbank* market is, in many ways, a gentleman's market. Dealers rarely renege on trades—not because they are paragons of virtue, but because to do so would certainly alienate others on the system. The last thing a dealer can afford to do is acquire a reputation as an unreliable counterparty; in such a case, the word gets around quickly and it becomes very difficult to find others with whom to trade. Also, dealers are expected to provide both a bid *and* offer when a counterparty asks for a price. Although the dealer may not want to show both sides of the market—perhaps his portfolio is already too heavily weighted to one side—he knows that he can expect to receive the same service from other dealers when the time comes to trade out of his portfolio.

One thing to remember about banks is that, no matter how big they and their credit lines are, they are not much different from the rest of us when it comes to trading. Banks hate to lose money, and they especially hate to assume risks that can't quickly be laid off onto someone else. In fact, one reason the turnover in FX is so incredibly high is because the banks pass around trades like a hot potato: a hedge fund puts an order in to sell $100 million Yen; a dealer buys the whole lot; she immediately sells $20 million lots to five dealers on EBS; each of those dealers, in turn, sells his or her portions to five new dealers; the circle has now widened to 25 additional counter-parties. This activity—some traders call it *velocity of trade*—is a never-ending circle of bids and offers, purchases and sales.

What's important to realize when trading FX is that one should not be intimidated by the banks. I remember visiting a dealing room in New York, in one of the largest currency trading houses in the world, a few years after I began trading. The trading floor must have been the size of a football field, and it was filled with what seemed like hundreds of smart, aggressive traders. I looked around at the state-of-the-art computer equipment, thought about the miles of phone lines snaking under the floor and connecting to other traders all over the world and the billions of dollars of credit lines supporting the traders' activity, and I began to wonder, "How can I possibly compete with this?" Then I overheard one of the traders whining about losing a trivial amount of money and I realized that even with all the tools available to them, they still made mistakes. The challenge for anyone who wants to trade the foreign exchange markets is to figure out how to capitalize on the errors that the big boys make. When you understand how the banks transact their business, it becomes reasonable to think that you can learn to beat them to the profitable trades.

INTERBANK PRICING CONVENTIONS

While many in the United States tend to express currency prices in dollar terms, this is contrary to the pricing convention in the *Interbank* market. There, quotes are expressed to reflect the amount of currency equal to one U.S. dollar. The quotes are reciprocals of each other. Let's say that the "dollar" price of the Canadian Dollar is 6660. In order to determine the equivalent *Interbank* price, divide 1/6660 and then move the decimal points four spaces to the right. The result is 1.5015015; however, *Interbank* prices are rounded to the fourth decimal place, so the quote becomes 1.5015. Obviously, since the cash and futures prices are reciprocals, it is simple to figure out a futures price from an *Interbank* quote. Simply follow the steps previously listed, in this case substituting 1.5015 for 6660.

SPOT AND FORWARDS

In the previous example, 1.5015 is the spot price, or today's value for Canadian Dollars. This does not mean that an FX dealer will sell you Canadian Dollars at spot. The dealer is in business to make a profit and will only agree to buy the currency at a discount to spot or sell it at a premium. For example, if a customer wants a two-sided "spot" market for 10 million Canadian Dollars, the dealer might bid 1.5012 and offer at 1.5018. By doing so, she is able to get an edge in relation to the spot price whether the customer buys or sells (see table below).

	Amount of Canadian Dollars	Value in U.S. Dollars	$ Difference to Spot
Dealer Bid (1.5012)	10 million	$15,012,000	+$3,000
SPOT (1.5015)	10 million	$15,015,000	
Dealer Offer (1.5018)	10 million	$15,018,000	+$3,000

Spot is only one type of transaction available in the *Interbank* market. A significant amount of business is done on a *forward* basis. Forwards are similar in concept to currency futures; they are contracts in which two parties agree to transact a specified amount of currency for delivery at a date in the future. There are, however, a number of major differences between futures and forwards.

Contract Size

Currency futures are for standardized amounts and are relatively small. The largest contract at the CME, for instance, is the Euro FX, which is worth approximately $110,000. By contrast, forwards trade in increments of at least $1 million. (Occasionally a customer will ask for a market for less than this amount and the dealer will usually accommodate the request. Since it is difficult, however, for the dealer to lay off such a small trade, he will tend to quote a wider spread to compensate for the additional risk and headache of carrying the unwanted position.)

Settlement

Currency futures are settled on four dates in the year: the third Monday of March, June, September, and December. Forwards, by contrast, are settled from three days to many years in the future. While the most common "even-dated" forwards are for 30, 60, 90, and 180 days, dealers may also customize a forward contract to the exact delivery date needed by the customer. In actual practice, the vast majority of forward deals are for 180 days or less. For the dealer, the risk involved in taking on a long-term forward exposure requires her to quote extremely wide spreads. Oftentimes the value of doing a hedge—and anyone who wants to buy or sell a 5-year forward is hardly looking for the next big pop in the market—is mitigated because the dealer's high price makes buying the protection too costly.

Pre-Settlement Liquidation

Just as one can liquidate a Japanese Yen contract at the CME prior to delivery, one can close out a forward position by entering into an opposite transaction for the same amount, to be delivered on the same date as called for in the original forward contract. There is, however, one major difference between futures and forwards with respect to early liquidation. In the futures markets, which are *marked-to-the-market,* the disbursement of funds takes place prior to the next trading session; liquidation is effectively the same as settlement. With a forward, although the trade has been liquidated, it is not settled. Both the original and offsetting positions remain open until the date the contract comes due. This can create difficulties in that both positions will count against the customer's credit lines until settlement even though there is no market exposure on the liquidated position. Accordingly, he may find himself denied sufficient credit to do subsequent deals in the interim.

Distribution of Gains and Losses

Both futures and forwards are marked-to-the-market, with gains and losses credited or debited in the customer's account every day. Just as with performance bonds (margins) in the futures markets, the customer must deposit additional funds to support her forward position if losses are greater than the amount of credit the dealer is willing to extend. The difference arises with respect to distributions. In a futures account, a customer can withdraw gains at any time; even unrealized gains, as long as sufficient capital remains in the account to meet the margin maintenance requirements.

Transaction Costs

In the futures markets, FCMs charge customers a per-contract fee. The range is quite wide: a relatively inactive retail trader may pay $20 to $50; an money manager, who is active and trades large size positions, may pay less than $6; a trader on the floor who owns a membership probably pays no more than 10 cents. In the forward market, there is no customer commission per se; instead, the dealer embeds his profit margin into the quoted spread.

Default Risk

While one should never say never when talking about risk, with currency futures there is almost no chance for the customer to suffer from a default of another market participant. The reason for this is that the CME clearing house, which is backed by its FCMs, guarantees every trade that is done on the exchange. Forwards, however, do entail some amount of risk in that there is no clearing house to guarantee performance. As is the nature of all private agreements, in the event that one of the parties to the deal defaults, the other will be forced to get in line with all of the other creditors.

Regulatory Oversight

Currency futures fall under the jurisdiction of the CFTC. The markets are heavily regulated and exchange members must follow a number of rules written specifically to protect the customer. The *Interbank* market is essentially unregulated in that it includes traders from virtually every country in the world. Because there is no body with jurisdiction over such a diverse group, it operates without formal rules such as those the CFTC imposes on exchange-based FX trading. In a sense, the *Interbank* is self-regulated; there are conventions and accepted business practices that any one involved must follow in order to join the club. Nonetheless, the lack of regulation is one of the key characteristics that distinguishes futures and forwards.

FOREX SWAPS

Generally, it is unusual for a customer to buy or sell a forward outright. If she wants to speculate in a currency, she is better off using futures or the spot market; after all, the dealer's tightest bid/ask spread will always be in spot since it is less risky than a longer-dated trade. Forwards are more often used in order to accomplish a *calendar spread trade* (Long one duration of an instrument versus another. For example, long the 30-day Yen forward/short the 90 day.) This type of trade is also known as a *swap*.

This is how it works: Let's say a banker needs to buy Canadian Dollars and it happens that he needs the currency for a month only. He will ask a dealer to make him a simultaneous market to sell Canadian Dollars at the spot price and buy them back at the 30-day forward price. If the price is to the bank's liking, it will enter into the agreement. These types of calendar trades can be entered into for just about any duration of time between the two instruments. The most common, however, are *spot/next* (exchanges the respective currencies in two days and reverses the trade on the third), *tom/next* (exchanges the currencies the next day and reverses the day after), and a *forward/forward* (one forward against a later-dated forward.)

It is not particularly important that you understand the technical aspects of how these trades are done, although if you are interested there are some excellent books on the topic which cover the mechanics and math behind *Interbank* trading (One good book on the subject, which can be understood even without an advanced degree in mathematics, is *Currency Risk Management* by Gary Shoup, Center for Futures Education, Inc., 1995.) What is important for you to know is that there is a giant world of FOREX trading out there. It includes thousands of traders in the *Interbank* and the futures exchanges; each with their own needs, ideas about the market, and trading strategies. This can seem frightening, especially to new traders. How, you may ask yourself, can I compete against these sophisticated players? Well, just remember that all of those sophisticated players are relying—or as traders say, leaning—on someone else in the market. When a dealer makes a two-sided price she first checks her trading screen to see where she can get out. When she's asked for a far-dated forward, she often makes a price so wide that the counterparty has to turn down the deal. When she loses money on a trade, her boss gets upset, her colleagues make fun of her, and she wonders how much longer she can continue to trade when she's such an idiot. In short, the sophisticated player is not much different from the rest of us.

MACROECONOMIC FACTORS THAT AFFECT CURRENCY VALUES

The actions of governments and central banks are the chief macroeconomic factors that affect currency values. Traders should carefully monitor the following conditions in order to help determine how to understand a currency's relative value:

- *Government Budget Deficits or Surpluses.* The market usually reacts negatively to widening government budget deficits and positively to narrowing deficits. The impact is reflected in the value of a country's currency.

- *Balance of Trade Levels and Trends.* The trade flow between countries illustrates the demand for goods and services, which in turn indicates demand for a country's currency to conduct trade.
- *Inflation Levels and Trends.* Typically, a currency will lose value if there is a high level of inflation in the country or if inflation levels are perceived to be rising. This is because inflation erodes purchasing power, and thus demand, for that particular currency.
- *Economic Growth and Health.* Reports such as the Gross Domestic Product (GDP), employment levels, retail sales, capacity utilization, and others detail the levels of a country's economic growth and health. Generally, the more healthy and robust a country's economy, the better its currency will perform and the more demand there will be for it.

TRADING CURRENCIES ELECTRONICALLY

Because few of us have the financial wherewithal or need to trade in the *Interbank* market, we will focus on opportunities in the futures markets. As anyone who has paid attention to currency trading in the last decade can tell you, the heyday of the CME's currency pits is gone forever. The list of reasons why this happened is long and probably an interesting subject for another book. For our purposes, it is sufficient to note that the currency pits failed to meet the basic expectations of the customers and so they took their business elsewhere.

In April of 2001, however, the CME took action to try to regain its status as an important marketplace by allowing its currency futures to trade electronically on a side-by-side basis with the contracts in the pit. Prior to the change, while one could trade electronically in the hours when the pits were closed if one wanted, they were prohibited from trading electronically during U.S. trading hours. This is especially incredible when one considers that electronic currency trading, unlike e-mini or T-bond trading, is no revolutionary concept; as we learned, it has been well-accepted since 1992. In fact, until the change it was probably fair to say that aside from a tiny bit of currency business transacted over the phone here and there and a small Dollar Index contract traded at the New York Board of Trade, the CME was the only currency marketplace in the world that *was not* electronic.

This side-by-side experiment is an attempt by the CME to offer the best of both worlds. To the extent that there are some customers who still appreciate the price discovery function of the pit, it is available to them. For everyone else, who expects to transact currency business electronically, the exchange is now a viable option.

THE GLOBEX FOREIGN EXCHANGE FACILITY

In 1996, in order to support its electronic currency business and ensure that there is always liquidity on GLOBEX, the CME created the GLOBEX Foreign Exchange Facility (GFX). The GFX is a unique undertaking in the world of foreign exchange trading. It is a currency dealing room with a staff of more than 20 traders and risk managers who makes markets in all of the CME currencies 24 hours per day. If the GFX were just another FX market-maker there would be nothing particularly noteworthy about it. It is different, however, in that it does not attempt to make a profit on its market-making activities. Its goal is simply to create liquidity so that CME members and customers can transact their business at the best possible price.

Here is how the GFX operates: It searches for the best FX price available in the cash market and translates it into a futures price. It then enters that price onto GLOBEX for a defined number of futures contracts and waits to get hit. In the event that there are resting GLOBEX orders that can be filled based on the translated price, the GFX will act aggressively to execute the orders. When the GFX makes a futures trade in either of the two scenarios previously explained, it *immediately* covers the exposure with a corresponding trade in the cash market. At that point, appropriately hedged, the GFX books the trade and at a later time closes out both the cash and futures portions to flatten the position. In most cases, the end result of these transactions is a break-even trade.

While the existence of the GFX is well-known to the relatively few individuals and entities that trade currencies on GLOBEX, the word has yet to spread to the wider world of currency traders. As the CME builds its electronic FX franchise, the GFX is unlikely to remain a secret because the benefit to the end-user is so profound. First, to the extent that the GFX is able to find liquidity in the cash market, it can effectively bring EBS and Reuters pricing to retail traders who would otherwise be restricted from those systems. Second, since the GFX has no imperative to make a profit—in fact, on occasions when it does make a profit on a trade it is acting outside of its mandate—it can bring wholesale currency pricing to the marketplace. Think about what an advantage this presents to the end-user. Other market-makers, be they banks, brokerage firms, or FX trading websites, will mark up their bid/ask spread just as a retail store marks up its goods and services in order to make a profit. With the GFX, however, there is no profit motive and, therefore, no need to charge a premium price. GFX approaches pricing its currencies the same way a convenience store prices milk: as a loss leader. In the same way 7-11 is prepared to lose money on every gallon in order to attract people into the store, the exchange is willing to underwrite an expensive and potentially risky endeavor in currency market-making to ensure that customers bring their business to GLOBEX. While much of the retail electric trade has gravitated towards e-minis, don't make the mistake of ignoring the CME currency contracts. With easy access to GLOBEX and the steadying presence of the GFX, CME currency futures offer great opportunities to the electronic trader.

8
TREASURY FUTURES

THE NEED FOR TREASURY FUTURES[1]

The Chicago Board of Trade (CBOT) Treasury Bond futures contract is one of the most actively traded contracts in the world. The T-Bond contract is the cornerstone of the CBOT Treasury Complex, which includes 10-Year Note futures, 5-Year Note futures, 2-Year Note futures, and options on these contracts. This group of trading products is vital to institutions and individuals across the globe, because it offers such widespread interest rate applications. There are six key reasons for the success of CBOT Treasury contracts:

- First, the increase in interest rate volatility worldwide creates a need for an effective, cost-efficient way to manage interest rate risk. International concerns about the relative strength or weakness of the U.S. dollar, changes in the U.S. Federal Reserve Board policies, debt and credit crises, and shifts in economic strength throughout the world have combined with other factors to make interest rate changes sharper and more dramatic. Volatility is now the rule rather than the exception.

---

[1] This material appears in *The CBOT® Financial Instruments Guide,* © Board of Trade, City of Chicago 1997-2001.

- Second, long-term financial assets such as U.S. Treasury bonds and notes are acutely sensitive to interest rate changes. Therefore, T-Bonds and T-Notes, as well as futures contracts on those issues, will respond more dramatically to interest rate changes than other instruments with shorter maturities.
- Third, U.S government securities have been issued and sold in record amounts, whetting the appetite of foreign investors and making U.S. Treasury Bonds a "global" commodity. U.S. Treasury rates often serve as benchmarks for interest rates on bonds issued by other governments.
- Fourth, U.S. Treasury issues are marketable in an active secondary market that knows no geographical or time boundaries.
- Fifth, because of the tremendous trading volume of CBOT Treasury futures, buying or selling is relatively easy and not disruptive to the market.
- Finally, there is the growing realization among market professionals that not using the futures markets is akin to speculating, because no one is consistently able to predict the magnitude or direction of interest rate changes over time.

THE BASICS OF CBOT TREASURY FUTURES

The specific commodity for CBOT Treasury futures contracts—the underlying instruments on which these contracts are based—is a $100,000 face value U.S. Treasury security for T-Bonds as well as for 10-Year and 5-Year T-Notes. Since the U.S. government issues significantly more debt in the 2-Year maturity sector than any other, there are more 2-Year securities traded in the underlying cash market. To accommodate the resulting need for trades of greater size, the CBOT designed its 2-Year T-Note contract to have a face value of $200,000 to provide economies of scale for Treasury futures market participants. The following chart illustrates the key features of the contracts in the CBOT Treasury complex.

The standardized dates of CBOT Treasury futures contracts call for delivery of the underlying security during the months of March, June, September, and December. Months listed for trading can extend more than two years into the future. Although futures contracts are based on the contractual obligation of delivery, only a small percentage of CBOT contracts are actually held until delivery. Because delivery plays such an integral role in the pricing of CBOT Treasury futures contracts, however, a clear understanding of the delivery process is essential.

	T-Bonds	10-Year Notes	5-Year Notes	2-Year Notes
Face Amount	$100,000	$100,000	$100,000	$200,000
Maturity	15 Years +	6 1/2 to 10 Years	4 1/6 to 5 1/4 Yrs.	1 3/4 to 2 Yrs.
Price Basis	6% Coupon	6% Coupon	6% Coupon	6% Coupon
Tick Size	32nds	32nds	32nds and 1/2 32nds	32nds and 1/4 32nds
Tick Value	$31.25	$31.25	$31.25	$62.50

Maturity and Delivery

CBOT Treasury Bond and Note futures contracts each have a range of cash issues that are eligible for delivery. A particular Treasury Bond or Note is eligible for delivery into the correlating futures contract as long as it meets the specified maturity requirements. It is the choice of the seller to determine which issue to deliver, and logic dictates that the seller would choose to deliver the cash instrument that is most economical. It is this cash instrument that the futures price tracks most closely. This advantageously priced instrument for delivery is known as the cheapest to delivery (CTD).

Pricing Basis

Since the obligations of a Treasury futures contract permit delivery of any U.S. Treasury security that meets the maturity requirements stated in its contract specifications, that contract may track any one of a number of bonds or notes eligible for delivery. Each of these may have different coupon rates. These securities may have different maturity dates as long as these are within the range described in the contract specifications. As a result, all may trade at different prices and at somewhat different yields to maturity. How then do the individual prices of these different instruments relate to a single price for a standardized futures contract?

To allow the futures price to reflect the full range of issues eligible for delivery, the CBOT developed a conversion factor system. This system was created to facilitate the T-Bond and T-Note delivery mechanism and adjust for the fact that there is a broad range of issues, with a broad range of coupons, eligible for delivery into the futures contract.

Conversion factors establish a correlation between the different prices of many eligible cash instruments and the single price of the corresponding standardized futures contract. A specific conversion factor is assigned to each cash instrument that meets the maturity specifications of a Treasury futures contract. This is used to adjust the price of a deliverable bond or note, given its specific maturity and yield characteristics, to the equivalent price for a 6 percent coupon.

For any delivery month, each deliverable issue has a specific conversion factor that reflects its dollar value and remaining time to maturity as of that delivery month. The easiest way to understand the conversion factors is to compare the coupon and maturity of the cash bond to the 6 percent coupon standard of the T-Bond futures contract.

Price Increments and Their Value

The Treasury futures market follows the conventions of the underlying cash market in quoting futures contract prices in points and increments of a point. A point equals 1 percent of the total face value of a security. Since futures on Treasury Bonds and 5- and 10-Year Notes are all contracts with a $100,000 face value, the value of a full point is $1,000 for each of these contracts. A one-point move on a $200,000 face value 2-Year T-Note futures would then have a value of $2,000.

Price movements for Treasury futures are denominated in fractions of a full percentage point. These minimum price increments are called *ticks*. T-Bond futures trade in minimum increments of 1/32 of a point, which is equal to $31.25. The 10-Year and 5-Year T-Note futures contracts trade in minimum price increments of 1/2 of 1/32, so the value of a *tick* in these contracts is $15.625. With the 2-Year T-Note contract double the size of the others and a minimum *tick* of 1/4 of 1/32, the *tick* value is $15.625 also.

Why are the *ticks* for the 10-, 5-, and 2-Year Treasury futures described as 1/2 and 1/4 1/32, rather than as 1/64 and 1/128? These are the conventions upon which the underlying markets are based. Also, the contracts that make up the CBOT Treasury complex must trade in a consistent pricing unit to minimize error and maximize the utility of these contracts. It would be too cumbersome while trading to remember to convert one price configuration to another among Treasury contracts.

For example, a price of 101-00 indicates that the futures contract in question is now priced at $101,000 (101 percent of $100,000). If the futures price increases to 101-16, this quotation must mean an increase of 16/32—regardless of which Treasury futures contract it represents. Thus, the price quote for 5-Year T-Note futures is given as 16/32 rather than 32/64 even though those values are equivalent. If the value of a 5-Year T-Note futures contract were to increase by 16 1/2 in the example above, its price would be 101-165, which means 101-16.5/32.

By the same logic, if a 2-Year T-Note futures contract priced at 101-00 increased in value by 16 1/2, its price would also be quoted as 101-165. Note the decimal point has been eliminated. This was done to accommodate price displays and software programs already in use before the introduction of these contracts. Quotes of 1/4 of 1/32 eliminate the last number in the field as well as the decimal point, again requiring fewer fields for its display. A 2-Year T-Note priced at 101-00 whose priced increased by 16 1/4 would be quoted as 101-162, meaning 101-16.25/32; a price increase of 16 3/4 would read 101-167, which is 101-16.75/32.

Note that the value of a *tick* is constant regardless of whether prices go up or down. A price movement of 24 *ticks,* for example, would change the contract value by $750 (24 × $31.25 per *tick*) on all Treasury contracts except for the 2-Year, where it will have double the value (24 × $62.50). If a 10-Year T-Note futures contract is priced at 99-00 and its price moves to 99-24, its value has increased by $750.

THE SIGNIFICANCE OF THE CHEAPEST TO DELIVER

Through the conversion factor system, a common point is established between the cash price and its equivalent futures price. This is critical in determining the cheapest to deliver instrument.

To Adjust a Futures Price to an Equivalent Cash Price

Adjusted Futures Price= Futures Price × Conversion Factor (or Cash Equivalent Price)

To Adjust a Cash Price to an Equivalent Futures Price

Adjusted Cash Price = Cash Price / Conversion Factor

Throughout its life, a futures contract tends to track the adjusted cash price of the cheapest to deliver (CTD) instrument; however, there still will be a price differential between the two. This price differential is known as the basis. Even though few futures contracts end in delivery, remember that a futures contract will most closely track the price movement of its underlying CTD security. This means that identifying the CTD is key to understanding how futures prices may be expected to move.

Basis and Carry

The "cheapness' of the CTD relative to the other securities eligible for delivery depends in part on carry. *Carry* is the relationship between the coupon income enjoyed by the owner of the security and the costs incurred to finance the purchase of that security until the future date when it is delivered into the futures market. Standard calculations in the cash and futures bond markets assume that financing is accomplished at the short-term repo rate. The *repo* (short form of repurchase) market may be thought of as a market for short-term loans collateralized by U.S. Treasury securities.

A simplified example might help to clarify the concept of carry. Suppose an investor wishes to purchase a Treasury Bond or Note with a 10 percent coupon. He finances the purchase by going to the repo market, where the security is used as collateral to borrow against its current market value. If the general level of the overnight repo rates is 6 percent and remains unchanged, he will earn 4 percent each day that he owns that 10 percent security. In this example, the investor earned more interest income on his cash bond than he paid to finance it. That is, he enjoyed a net interest income, or positive carry. *Positive carry* is characteristic of an upwardly sloping yield curve, where long-term rates are higher than short-term rates. If there were an inverted yield curve, where short-term rates exceeded long-term rates, *negative carry* may result and the investor could spend more than he earns to finance the purchase of the cash instrument.

In the positive carry example, where an issue with a 10 percent coupon was financed at 6 percent, the buyer of the security will hold it until the time comes to deliver it into the futures market. But the buyer of a corresponding futures contract will not earn interest income because a futures contract is not an interest bearing instrument. If the price of a CTD cash market security were always equal to the price of its corresponding futures contract, then there would be an enormous advantage to buying cash market securities rather than futures contracts. To maintain a balance in the relationship between cash and futures prices, therefore, the price of a futures contract must be discounted by the dollar value of the carry on the underlying CTD security.

Logically, in a positive carry environment, the futures price should equal the adjusted price of the CTD instrument less the positive carry. In this case the carry benefits (interest income) for Treasury securities exceed financing costs (carrying charges). As a result, the price drops below adjusted cash market prices. Conversely, in a negative carry, or inverted yield curve environment, the futures price should equal the adjusted cash price *plus* carrying charges. As a result of these carry implications, the more deferred the expiration of a Treasury futures contract, the lower it should price relative to the cash market value of its underlying CTD security in a positive yield curve environment. And this price differential—the basis—should shrink as time elapses and the contract approaches expiration. This principle is known as conversion and implies that cash and adjusted futures prices will be approximately equal on the expiration date of a futures contract.

There is another component to basis, although it generally has less impact than carry on futures prices for the nearby expiration. The design of the Treasury futures contract gives the short the choice of what issue to deliver, as well as when to deliver (within the delivery window). These choices create what is called an *implied put option*, which favors the short. Of course, the long is compensated for the option benefiting the short. The market's assessment of the benefit of the short's delivery option is reflected in the actual delivery price. While carry may be either positive or negative, the value of the implied put option is always positive since there is always a value associated with the ability to make choices. Like positive carry, the value of the implied put option must be subtracted from the adjusted futures price. As a result, the futures contract frequently prices at a discount to the cash equivalent (adjusted futures) price of the CTD Bond or Note.

SOME REAL WORLD APPLICATIONS

Short Hedge

A corporate treasurer is in charge of her company's bond position. She has structured the portfolio to her satisfaction but is concerned about falling prices if next week's unemployment figures are bad. She is willing to exit the market, foregoing potential gains in order to avoid a potential loss, but liquidating her cash position is difficult and costly. Instead, she neutralizes her position by selling bond futures. After the release of the figures the bond market actually rallies a bit. The treasurer takes a small loss on her futures position, but it is offset by the gain in her portfolio. Despite losing money, the hedge serves its purpose: to neutralize her market exposure. Had futures prices fallen, the treasurer's hedge would have profited while her cash position would have suffered losses. The profit from the hedge would cancel out the portfolio's loss, and again she would break even.

A portfolio manager holds $1 million face value of the 8 1/8 percent U.S. Treasury Bonds maturing on May 15, 2021. The cash price is currently 120-00. The manager fears interest rates will rise sharply during the next few months and wants to protect the value of the bond from its expected decline in resale value and consequently decides to sell 10 T-Bond futures contracts, currently trading at 95-00.

Cash Now	Futures
Holds $1 million face of 8 1/8% of May 21 at 120-00, or $1.2 million	Sells 10 March Bond futures contracts at 95-00, or $95,000
Cash Later	*Futures*
Market Value of Treasury falls to 105-00, or $1,050,000 (Net change of − $150,000)	Buys 10 futures contracts at 83-16, or $835,000 (Net change of $115,000)

Result

Several weeks later, interest rates have risen. Although all but $35,000 of the portfolio manager's cash market loss was offset with the Treasury Bond futures, the results would have been improved by adjusting the number of futures contracts used to hedge the cash market holdings.

The Weighted Short Hedge

To compensate for the greater decline in the dollar value of a cash bond compared to the decline in the value of the nearby T-Bond futures contract, a weighted hedge is often used. One method of weighted hedging involves the use of conversion factors.

The portfolio manager in the previous example hedged $1 million face value 8 1/8 percent U.S. Treasury Bonds maturing May 2021 by selling 10 March T-Bond futures contracts. The result of this hedge could have been improved if the manager had adjusted the number of futures contracts used. To determine the number of T-Bond futures needed for delivery against the March 2000 futures contract, the portfolio manager would make the following calculation:

Number of
Futures Contracts = Conversion Factor × (Par Value of / Par Value of
                                         Cash Bond   Futures)
= 1.2518 × ($1 million/$100,000)
= 1.2518 × 10
= 12.518 Futures Contracts

Since it is not possible to trade a fraction of a futures contract, the manager rounds up to the nearest whole number, in this case 13.

Assuming the previous situation, the results obtained by using the weighted hedge are considerably different.

Cash Now	Futures
Holds $1 million face of 8 1/8% of May 21 at 120-00, or $1.2 million	Sells 13 March T-Bond futures contracts at 95-00, or $1,235,000

Cash Later	Futures
Market value of Treasury falls to 105-00, or $1,050,000 (Net change of −$150,000)	Buys 13 futures contracts at 83-16, or $1,085,000 (Net change of $146,250)

Result

In this instance, the advantage of using the weighted hedge is clear. The portfolio manager has come close to achieving full offset of this risk.

The Long Hedge (Anticipatory Hedge)

Suppose bonds are trading at relatively cheap prices and a pension fund manager wants to buy some for her portfolio. Unfortunately, she must wait for funding which she expects three months from now. The fund manager worries that by waiting she will miss out on low prices. So, to participate in the market, she buys bond futures. After three months she receives her funding. As feared, bond prices do climb higher, and she has to pay the more expensive price. But futures prices trade higher too, and the fund manager realizes a profit on her futures position. She sells out her futures, applies the profits toward her cash purchase, and effectively lowers the net purchase price of the bonds.

Consider the following scenario: A corporation has a $100,000 face amount U.S. Treasury Note in its portfolio that will mature in several weeks. The corporate treasurer knows that he will reinvest the money to be received by buying a 10-Year Treasury Note issued earlier in the year. He is afraid that 10-year interest rates may fall before the maturity date arrives. If they do, he will then have to pay a higher price for the note that he now plans to buy. To offset such a price increase, he buys a 10-Year Note futures contract.

Cash Now

10-Year Cash T-Note priced at 98-16 or $98,500

Futures

Buys 1 10-Year Note futures at 102-00, or $102,000

Cash Later

10-Year Cash T-Note priced at 99-00, or $99,000

Futures

Sells 1 10-Year Note futures at 102-16, or $102,500

Result

Interest rates did fall before the corporation could afford to buy the 10-Year T-Note and so the price of the $100,000 10-Year Note increased. The treasurer pays $500 more to buy the Note than he would have several weeks earlier. Meanwhile, the futures position gained $500, which offset the increased cost of buying at a higher price. In effect, the treasurer locked in the cheaper purchase price by buying Treasury futures.

As was discussed earlier, any change in the yield of a coupon-bearing security produces a simultaneous change in its price, and the direction of this price movement will be inverse to the direction of the interest rate movement. The magnitude of the price change resulting from a given change in the yield of a bond or note depends on a variety of factors: its coupon rate, the proximity of that rate to prevailing yield levels, and—most significantly—its remaining time to maturity. Price sensitivity will vary among U.S. Treasury issues because of differences in their stated coupon rates and maturities.

Although cash and futures prices tend to move in roughly parallel fashion, they don't always do so at the same rate. Thus, simply matching futures contract amounts with cash market amounts may not always produce the most favorable results. Take Treasury Bonds, for example. If all other factors were held constant, the prices of two bonds with identical coupon rates and maturities would move together in near perfect harmony. Matching $1 million in standardized T-Bond futures contracts against a cash market position of $1 million in cash market bonds may not always provide optimal protection if the cash bond is different from the 6 percent coupon standard of the T-Bond futures contract. To the extent that price movements between futures and cash prices may not correspond perfectly with each other (basis risk), the futures hedge position may generate some surplus profit or loss.

The Weighted Long Hedge

A pension fund manager wants to take advantage of the current yield of the 10-Year T-Note but won't have the necessary cash to do so until a few weeks later, when she plans to buy $10 million face amount of the 10-Year T-Note. The 10-Year Note she intends to buy has a 6 5/8 percent coupon, matures in May 2007, and is priced 103-02+. The conversion factor for this note is for the March 2000 10-Year T-Note futures delivery month is 1.0353 and the March futures contract is trading at 98-11+ (11+ indicating 11 5/32).

Using a weighted hedge, she goes long 104 contracts [1.0353 × ($10 million/$100,000)]. When she is ready to make the actual Treasury purchase, these futures are trading at 99-07.

Cash Now	Futures
$10 million face of T-Note at 103-02+, or $10,307,810	Buys 104 March 10-Year Note futures contracts at 98-11+, or $10,229,375
Cash Later	Futures
$10 million face of T-Note at 104-00+, or $10,401,830 (Net change of $94,020)	Sells 104 March futures at 99-07, or $10,318,750 (Net change of $89,375)

Result

Use of the long hedge produced measurable benefits. Had the long hedge not been used, it would have cost the manager an additional $89,375 to acquire the T-Note in the cash market. By using the futures to hedge, the manager considerably reduced the adverse effect of the 15 basis point yield drop.

Spreading

Spreading involves the simultaneous purchase of one futures contract and the sales of another, with the expectation that their price relationship will change enough to produce a profit when both positions are closed. Spreads are popular for two reasons—less risk and lower cost. Because spreads are generally less risky than an outright long or short position, exchange set margins are generally lower.

Many Treasury spread strategies are intended to take advantage of the differences in price sensitivity between similar instruments of different maturities. With all else being equal, the greater the length of time until maturity, the greater an impact a change in yield will have on price. The price of a T-Bond futures contract, therefore, should move more dramatically in response to a given change in yield than the prices of the 2-year, 5-Year, and 10-Year T-Notes.

Example

An investor anticipates an increase in interest rates during the next several weeks. Since he expects rates to rise similarly across all maturities, he expects long-term prices to decline more than intermediate term prices. Therefore, he buys 10-Year T-Note futures at 102-00 and sells T-Bond futures at 98-24. Three weeks later, T-Bond futures are at 95-00 and T-Note futures are at 100-00.

10-Year T-Note Futures Now	T-Bond Futures
Buys 1 contract at 102-00, or $102,000	Sells 1 contract at 98-24, or $98,750
10-Year T-Note Later	T-Bond Futures
Sells 1 contract at 100-00, or $100,000 (Net change of −2-00, or −$2,000)	Buys 1 contract at 95-00, or $95,000 (Net change of 3-24, or $3,750)

Result

The net profit to the trader from the spread trade was 1 24/32, or $1,750.

THE FEDERAL RESERVE

The Federal Reserve (Fed) sets targets for short-term interest rate levels. To reach its objectives, the Fed has three tools at its disposal:

1. Buying or selling government securities through open market operations
2. Raising or lowering the discount rate
3. Raising or lowering the reserve requirements of member banks

Open Market Operations

The buying and selling of government securities (bills, notes, and bonds) by the Federal Reserve are known as open market operations, because these operations are conducted publicly. When the Fed sells securities it takes money out of the banking system. This "tightens" the money supply, which in turn puts upward pressure on interest rates. When the Fed buys securities, however, it injects money into the banking system to reduce pressure on interest rates.

Fed policy on interest rates and the means chosen to implement those policies have changed over the years. Throughout the 1970s, the Fed was concerned with targeting interest rate levels. For the next decade, daily Fed activities were directed toward controlling the growth of the money supply and inflation. There was no disclosure by the Fed about the levels targeted for the money supply and the intended effect of its activities on interest rates. Therefore, the market scrutinized Fed open market operations, interpreting these to discover the Fed's underlying intent.

In the 1990s, however, interest rate levels again became the focal point of Fed action. Following this, a profound change took place in the way the Fed signals its decision regarding monetary policy. The Fed began a new policy of immediately disclosing its decisions to the marketplace, following the February 1994 meeting of the Federal Open Market Committee (FOMC). While open market operations continue to be the tools used to implement its policies, the Fed now announces its target levels, and the market may expect the Fed to conduct whatever activities are needed daily to reach those levels. Most announcements concerning its target rate are now made during regularly scheduled FOMC meetings. Although the Fed is free to conduct open market operations at any time during a business day, it usually does so around 10:40 a.m., Chicago time.

The Discount Rate

The discount rate is the rate of interest the Fed charges on money borrowed by member banks. When the discount rate is lowered, banks have an incentive to use their borrowing privilege, increasing the amount of funds available to lend. In this instance, credit is "eased."

However, an increase in the discount rate makes the cost of money more expensive and credit is then "tightened." The discount rate can be lowered or raised as often as the Fed deems necessary.

Reserve Requirements

By law, banks are required to have a certain amount of "spare cash," or "reserves," on hand. If banks must keep a larger percentage of their assets as reserves, less money is available for consumers in the form of loans or mortgages. The Federal Reserve has the ability to require member banks to raise or lower reserves as conditions warrant.

Since reserves are not interest-bearing deposits, it is best for each bank to keep no more than the required minimum on deposit. Once these reserve requirements are satisfied, member banks then loan excess reserve funds, known as fed funds, to banks in need of additional cash. The rate of interest on fed funds is closely watched by the financial community, because it changes from moment to moment. The fed funds rate is a short-term interest rate that generally acts as a benchmark for other short-term rates.

Cause and Effect: The Federal Reserve Board Policy

The price direction of fixed-income instruments and their corresponding futures contract is impacted by information entering the marketplace. This includes Fed actions and economic announcements as well as certain key news events. The importance associated with any of these events depends on current economic conditions, so their impact on prices changes with time. Even then, conflicting information may make it difficult to predict the Fed's response. Fed policy objectives are achieved through the activities described below. If economic data indicates an accelerating economy, the Fed will act to slow it by taking steps to raise rates. The reverse takes place when the economy shows sign of decelerating. The Fed will act to stimulate growth by lowering rates. While the impact of the events listed below may change with market conditions, it may be useful in understanding the factors that influence market direction.

>Fed Raises Discount Rate . . . Market Goes Lower
>
>Fed Lowers Discount Rate . . . Market Goes Higher

Rationale: An increase in the borrowing rate for banks usually results in increased rates for the bank's customers. A decrease in the borrowing rate usually results in decreased customer rates.

>Money Supply Increases . . . Market Goes Lower

Rationale: Growth in the money supply in excess of population growth is usually inflationary. The Fed may then tighten monetary policy by raising the discount rate.

>Fed Does Repurchase Agreements . . . Market Goes Higher
>
>Fed Does Reverse or Matched Sales . . . Market Goes Lower

Rationale: If the Fed puts money into the system by purchasing collateral and agreeing to resell later, there is more money available for lending. In such a situation, rates should tend to move lower.

Matched sales have the exact opposite effect as the Fed essentially borrows the available liquidity in the system, forcing rates higher.

Fed Buys Bills . . . Market Goes Higher
Fed Sells Bills . . . Market Goes Lower

Rationale: In the first instance, the Fed lends money to the banking system, essentially increasing the supply of funds that are available for lending. An excess of funds will tend to drive interest rates lower. If the Fed sells bills, they decrease the supply of liquidity in the system and rates go up.

Causes and Effect: Key Economic Indicators

The causes and effects shown below are generally consistent with market experience; however, they are not guaranteed. The degree of impact caused by any one factor will change over time. Key market indicators in an inflationary environment, for instance, may be of little importance in a healthier economy. The impact on price direction may also shift as economic conditions change, so the information that now signals rising prices may one day cause them to drop. Furthermore, the market often receives conflicting information, in which case the direction of price response will reflect the dominant factor in the current environment.

Traders often anticipate key economic releases, estimate what the data will show, estimate its impact on price, and position themselves accordingly. If there is widespread agreement on these factors, the market will "price in" the data before its release. Should expectations prove correct, there will be little or no subsequent effect on price, and a casual observer might falsely conclude that the release had no impact on price at all. If, on the other hand, the data contradicts market assumptions, prices may move sharply. The direction and magnitude of these price corrections will now depend entirely on what assumptions had been priced into the market and on how different these were from the facts released. As a result, the direction of price movements often appears to be inconsistent with, or even contrary to, the broad guidelines listed below.

Consumer Price Index Rises . . . Market Falls
Consumer Price Index Falls . . . Market Rises

Rationale: CPI is a measure of inflation. Rising inflationary pressures tend to force interest rates higher. A lower CPI portends lower rates.

Durable Goods Orders Rise . . . Market Falls
Durable Goods Orders Fall . . . Market Rises

Rationale: Increasing business activity usually leads to increased credit demand which pushes rates higher. Less business activity means lower demand and lower rates.

Gross Domestic Product Rises . . . Market Falls
Gross Domestic Product Falls . . . Market Rises

Rationale: Accelerating economic activity may lead to higher demand for borrowing and prompt the Fed to increase rates. Slowing economic activity may cause the Fed to lower rates to stimulate the economy.

Housing Starts Rise . . . Market Falls
Housing Starts Fall . . . Market Rises

Rationale: Economic growth and increased demand for credit may force Fed to raise rates. If demand for funds decreases, the Fed may lower rates to stimulate the economy.

Industrial Production Rises . . . Market Falls
Industrial Production Falls . . . Market Rises

Rationale: Accelerating economic growth increases the likelihood that the Fed will increase rates. The Fed may let rates fall as economic activity slows.

Inventories Fall . . . Market Falls
Inventories Rise . . . Market Rises

Rationale: If inventories are down, it indicates an accelerating economy since sales are increasing faster than production. The market will decline because it will expect the Fed to raise rates to stave off inflation. When inventories rise it signals a slowing economy with sales lagging behind production. In such a situation the Fed may lower rates.

Leading Indicators Rise . . . Market Falls
Leading Indicators Fall . . . Market Rises

Rationale: Greater strength in the economy leading to increased credit demand may result in a Fed action to raise rates to slow down the economy. In a slowing economy, the Fed may lower rates.

Personal Income Rises . . . Market Falls

Personal Income Falls . . . Market Rises

Rationale: More income may prompt increased demand for consumer goods, resulting in higher prices. This inflationary pressure will tend to lead towards higher interest rates. Less income will have the opposite effect.

Producer Price Index Rises . . . Market Falls

Producer Price Index Falls . . . Market Rises

Rationale: Rising inflation may lead the Fed to increase rates. Lower inflationary pressures will lead to an opposite effect.

Retail Sales Rises . . . Market Falls

Retail Sales Falls . . . Market Rises

Rationale: Stronger economic growth may cause the Fed to raise rates to stave off inflation. Less economic growth will have an opposite effect.

Unemployment Rises . . . Market Rises

Unemployment Falls . . . Market Falls

Rationale: A weak economy increases the probability that the Fed will lower rates. An accelerating economy may have to be slowed down with higher rates.

9
BECOMING A SUCCESSFUL TRADER

DO YOU HAVE WHAT IT TAKES TO BECOME A SUCCESSFUL TRADER?

I started trading when I was 23 years old. I had worked during the previous four summers as a clerk on the floors of the CME and CBOT, and though I went through college telling everyone I was going to be a lawyer, I knew all along that I really wanted to be a trader. I did actually attend a semester of law school but hated it all: every case, and every Latin phrase. Instead of taking notes during class, I sat in the back of the room constructing charts and drawing trend lines. While I could not tell you what a habeas corpus was, I could quote whole sections of Edwards and Magee's *Technical Analysis of Stock Trends*. Though I should have been reading cases, I spent that entire semester devouring books about the markets: *Reminiscences of a Stock Operator* by Jesse Livermore, *A Random Walk Down Wall Street* by Burton Malkiel, and *Extraordinary Popular Delusions and the Madness of Crowds* by Charles MacKay. I studied fundamental analysis, discovering how ocean currents in South America affect grain prices in Chicago, how the Federal Reserve tries to manage interest rates to control inflation, and how the world's currencies react to the balance of trade between nations. I began to slip phrases like "supply and demand," "reserve requirements," and "yield curve" into conversations, exasperating my friends. I pontificated out of youthful exuberance and masked my inexperience with undeniable passion.

This passion I felt for the market was the foundation upon which I built my professional success. Some will tell you that trading is simply about making money: buying low and selling high; mechanically and without emotion. I guarantee you that those people are not making as much money as traders who are excited about what they do for a living. If you are not committed to becoming a serious student of the market, you are better off going to Vegas than opening a trading account. There, at least, you can catch a show before you lose all your money.

IT'S GOING TO TAKE SOME TIME

Occasionally, one will hear of, or meet, a trader who is a natural; someone who makes money from the instant he or she starts trading and cannot understand what the rest of us find so difficult about it. I know a few of these people, but they are exceptions to the rule. It is far more common to see novice traders struggling: struggling to understand the vicissitudes of the market, as well as struggling to overcome their fears and inhibitions. For most, trading is an evolutionary process. You begin as a weak, defenseless organism. Your puny brain seems to respond to stimuli in slow motion and you do stupid things like adding to losing positions, fading the market, and promising God to "never make the same mistake again, if He will just save you this one time." Predators lurk and entice you into their traps. If, somehow, you survive, you may earn a place on the food chain, but you cannot rely on your past successes to keep you there. With every *tick* of the market, others will try to remove you. You must constantly adapt or give way to your competitors. Such is the process of natural selection in the marketplace.

The first goal must be to learn how to survive the initial stage of development. No two traders are alike, so it is difficult to say how long one should expect it will take to start making money. You should be aware that it is highly unusual for novice traders to make money right away. (In the rare cases when this happens, it can be the worst possible turn of events because it can skew one's expectations.) As a general rule, it is unreasonable to expect that one can be making money consistently until spending between three to six months trading. Interestingly, professional floor traders who make the switch to electronic trading often find that their learning curve is far steeper than that of someone who is brand new to the markets. Although floor traders have valuable market experience, and in some cases substantial amounts of trading capital, many of the skills they have developed throughout their years on the floor are no longer useful, and in some ways are inimical, when trading electronically. They can no longer rely on the cues of sight and sound and the relationships they have built during their years in the pit.

Irrespective of whether one has trading experience, it is extremely important to recognize that it will take time to become successful when learning to trade electronically. Think about any endeavor with which you've struggled to develop an expertise—playing an instrument, learning how to shoot a jump shot, or obtaining an academic degree—and remember all the hours, weeks, months, or years it took to become proficient. It is hard to understand why anyone would think that becoming a good trader requires any less of an effort.

Throughout the learning process, as difficult as it may be, you should entirely forget about making money. It sounds somewhat counter-intuitive and easy for me to say, but instead of focusing on profitability during this period your goal should be to learn something new every single day. In fact, you should try to learn something new with every single transaction. Take a systematic approach to learning the fundamental trading techniques:

- Develop confidence in simple trade entry. Establish a goal of placing at least 5 to 10 orders per day into the market in order to learn how to use the keyboard and mouse. At the outset, you don't even need to make trades. Place orders off of the market so that you can see them enter the book. Then remove them from live action by cancelling them before they are executed.
- Watch your charts. Spend time learning how to identify patterns, particularly trends. Understand the difference between long- and short-term trends. Look for patterns that signal high-probability trades. Start to enter orders based on your understanding of the charts.
- Begin to appreciate the natural ebb and flow of orders during the trading session. Concentrate on how the market behaves on the openings and closes and when it slows down during lunch time. Watch what happens during fast markets and after you have seen a number of them find the courage to enter an order.

- After you are comfortable using the keyboard and mouse to enter trades, establish a goal of making at least 5 to 10 trades per day. Try to buy on the bid and sell on the offer. If you are hit, look immediately to close out the position: if you buy at 17, offer at 18; If the market goes to a 17 offer, offer 17 but get ready to take a loss by hitting the 16 bid if necessary. Once again, the amount of money you make or lose on these trades is relatively unimportant. Remember, at this stage, the goal is simply to learn how to make trades.
- After you have developed some confidence getting in and out of the market, as previously suggested, try something more aggressive. When the market starts to move in a particular direction, reach out and buy the offer or sell the bid for a single contract. Recognize that if the market is moving rapidly you must widen your exit parameters. Whereas in a normal market you might look to close out a position within a *tick* or two, in a more volatile move it may not be possible to do so.
- Try to spend as much time as possible in front of your PC concentrating on the market. Sometimes you will want to walk away—and it is important to take regular breaks—but ideally a new electronic trader should spend at least 80 percent of the trading day in front of the monitor, even if it is just to watch. Remember that every moment you are doing something else, other market participants are honing skills that they will use to compete against you.

The most important thing to remember during the first few months is that every day is an opportunity to learn something new: to develop confidence and move ahead of someone else who is not working as diligently as you are. At some point, you will realize that things are getting easier. You will stop looking at the chart with every *tick* because the pattern is embedded in your mind. Your volume will go up and an increasing percentage of trades will be winners. Although you will have losing trades, you will exit before they overwhelm you. Instead of dreading fast markets, you will look forward to them and complain when they are too few and far between. On busy days, you will find yourself skipping lunch and ignoring the call of nature because you will not want to miss a single *tick* of the market. You will start talking and thinking like a trader. And guess what? You will be making money.

THERE ARE NO SHORTCUTS

As you read through any futures industry publication you will find numerous advertisements, particularly towards the back cover, for all sorts of services and systems that promise windfall profits. One that appeared within the last year in *Active Trader* Magazine was "298 percent Return in 2000!" In the testimonial section—there are always testimonials in these ads—J.G. from Michigan (the initials have been changed to protect the gullible) writes, "This is a no-brainer system." I would like to think that the vendors of this and all the other no-brainer systems believe that they are providing the trading community with a valuable service. But when I read ads like these, the skeptic within me says these systems have to be "no-brainers" because the people who buy them . . . well, you get the idea. Am I being too harsh? Think about whether this sounds logical: Someone who is capable of turning a single dollar into $298 in just one year is willing to share his secrets with you if you will send him a few hundred dollars. A trader whose performance is better than Warren Buffet's is likely to have higher aspirations.

While it is easy to summarily dismiss these get-rich-quick schemes, the reality is that lots of people buy systems. Human beings are predictable that way; always looking for the easy way out. How many of us stayed up the night before an English midterm reading the Cliff's Notes on *Jane Eyre?* And don't you just hate those "three yards and a cloud of dust" offenses? Throw the ball 70 yards up field every down! In the spirit of full disclosure I admit that, as a very young and inexperienced trader, I purchased a system. It required me to fade the market if it closed in a particular direction for three days in a row. The theory was that after three days in one direction, the market would snap back. Unfortunately, the system did not contemplate that the grains might close limit-up eight days in a row with me short in everything from soybeans to Kansas City wheat. It was an expensive lesson, but I learned it well: I was not going to make it by trying to trade with "Cliff's Notes" as a guide. I also learned that the definition of a worthless system is one that is for sale.

Successful trading requires knowledge, concentration, and commitment, and none of those things can be bought. I am constantly amazed at the number of new traders I talk with who think that they can get by with less than a complete effort; that by working a few hours a week they can compete effectively against professional traders. In the movie *Spinal Tap,* one of the characters says that the band uses a special amplifier that doesn't stop at 10, it goes to 11. "It's 1 louder," he insists. Successful traders operate at 11 and, when necessary, can crank it up even higher. Someone once suggested to me that the source of this incredible misconception—that one can trade successfully without a full-time commitment—is that on the surface, trading appears to be easy. If, for example, you attempt to perform brain surgery, there is a 0 percent chance the patient will come out alive. If, on the other hand, you try to predict whether the price of soybeans will be higher or lower one week from today, there is a 50 percent chance you will be correct. While a 50/50 chance may make profitable trading seem attainable, after taking into account commissions, the bid/ask spread, margin requirements, the time value of money, and the trader's own fragile psyche, the odds of succeeding diminish significantly. Therefore, trading—while not as inherently important as brain surgery—does share a similar characteristic: the ability to do either successfully resides, as it should, in the hands of the professional.

The idea that anyone can make money trading is kind of like saying any child can grow up to be President. It is technically correct, but, of course, very few make it to the Oval Office. Trading for a living is hard, and someone who tries to tell you that anything about it is easy is lying or delusional. Consider this: there are individuals who try trading that make a serious commitment to achieve excellence and many of those people fail. How much more difficult will it be for someone who expects instant gratification? The lesson to be learned here is that if you are serious about making it as a trader, you can expect a difficult, time-consuming, and painful process. If you make it, however, the payoff is appropriate: you get to take away money from all of the unprepared and lazy traders who were not capable of reaching your level of proficiency.

DEVELOP A TRADING METHODOLOGY

Every successful trader I know trades with a plan. It may, in some cases, be systems-oriented with exact rules that must be followed. In most cases, however, traders adhere to less-rigid conventions. For instance, when I was a new trader a family friend who had been trading successfully for years took me out for lunch one day. We talked for a while about the market, and before long I began to complain about how

I could not seem to get the hang of things. He gave me a bit of a pep talk and then, looking over his shoulder as if concerned that someone would overhear him, he offered to tell me the trading rules—his self-described ten commandments—he had followed to build his fortune. They included such gems as, "Keep your losses small and let your winners run (STOP THE PRESSES!)." Although his "secrets" were mostly common sense and, even in those less sophisticated times, already clichés, I copied every word he said onto a trading card which I kept in the breast pocket of my trading jacket for many years. Clichés or not, I followed those rules like a zealot. I frequently referred to that card during those years—rereading it so many times that the ink faded under the force of a hundred thousand thumb prints—and was actually despondent when I came to work one day to find that the company that laundered my trading jacket had accidentally thrown out the contents of my pockets, including the treasured card. At that point, I did not really need it as anything more than a talisman. I had memorized the rules long before. More importantly, after all those years, the rules were second nature to me. This is what every trader should strive for: to develop a set of fundamental beliefs so deeply held that it becomes inconceivable to think of deviating from them.

THE TEN COMMANDMENTS

1. *A strong/weak market is one that trades:*
   - *above/below today's opening range*
   - *above/below yesterday's opening range*
   - *above/below yesterday's high/low*

   The opening range acts as a pivot point around which the market fluctuates until it decides which way it is heading. The high and low can be viewed as pivot points as well. They are even stronger indicators of strength/weakness than the opening range. Therefore, if the market is trading ABOVE today and yesterday's opening range as well as above yesterday's high, look to buy the market on breaks. If you must sell, do so cautiously, as the percentages are with the longs. Similarly, if the market is trading BELOW today and yesterday's opening range and below yesterday's low, look to sell the market on rallies. If you must buy, do so cautiously, as the percentages favor the seller.

2. *Volume drives the market higher/lower:*
   - *buy a rising market that goes higher on high volume*
   - *sell a rising market that goes higher on low volume*
   - *sell a declining market that goes lower on high volume*
   - *buy a declining market that goes lower on low volume*

   One of the best signals available to discern the staying power of a market move is volume. Do not fade a high-volume move. Consider fading a low-volume move.

3. *Whenever possible, buy on the bid and sell on the offer. Trading with the edge makes it easier to get out with a profit, without having to buy the low or sell the high of the move.*

4. *If it is hard to buy on the bid, keep trying to buy. If it is hard to sell on the offer, keep trying to sell. When a trade is hard to get, it is worth pursuing.*

5. *When you are trading well, trade your maximum position size; when you are trading poorly, cut your position size.*

   Stanley Druckenmiller, the great money manager, put it best in his "Market Wizards" interview (Jack Schwager, *Schwager on Futures,* John Wiley & Sons, 1996, page 759): "The way to build superior long-term returns is through preservation of capital and home runs. . . . When you have tremendous conviction on a trade, you have to go for the jugular. It takes courage to be a pig."

6. *Scale into and out of positions.*

   Putting your entire position size on at one time forces you to be perfect. You must use tight stops, perhaps tighter than the market warrants. In such cases it is extremely common to see traders get out with a loss, only to watch in frustration as the market moves in the original direction of their position. Had the trader not felt constrained by the size of the position, it would have been possible to stay with it long enough to appreciate the anticipated market move.

7. *You need to get paid for taking risk. Aim for a risk/reward ratio of 2:1 on your scalps and 4:1 on your position trades.*

8. *Get out when you can, not when you have to.*

Occasionally you will sell the high or buy the low of a move. There is no way to predict the high/low before the fact so the best we can hope for is to sell into strength and buy into weakness at saturation points. It is far better to get out too soon than too late.

9. *Adjust position size for volatility.*

   If, for example, the market is twice as volatile as normal, you should adjust your position size downward by a factor of one-half. Anyone who does not understand this basic principle should do everyone a favor and donate his entire trading account to charity. At least that way someone worthy will derive the benefit of your stupidity and you will have a nice tax deduction . . . not that you will need it.

10. *Don't hold your losers and never, never, never add to a losing position.*

    See above.

WHAT TYPE OF METHODOLOGY SHOULD I USE?

This is a difficult question to answer generically. Every trader has different wants, expectations, and tolerances. You will need to determine what type of approach suits you best and stick with it. There are certain common attributes in all successful approaches, but when it comes to the details there are as many ways to make money as there are money-making traders. One of the unshakable rules on my long-lost trading card was: "Don't hold your losers and never, never, never add to a losing position." Yet, I know great traders who, in a rapidly breaking market, buy a 20 lot at a price of "30," a 30 lot at "20," and for good measure, a 50 lot at "10" to improve their average price (in this case, to a total position of 100 lots at 17). Another of my rules is to never carry a position just prior to the release of an important economic figure. Others, however, look forward to those scenarios. They welcome the uncertainty of the moment and pray for volatility. I even know someone who makes trading decisions based on the phases of the moon. While I would sooner howl at the moon than makes trades based on whether it is full, I recognize that my methodologies are not necessarily superior to his. Plus, he lives in a much nicer house than I do. It almost does not matter what the methodology is, so long as the trader believes in it and makes money with it.

FIND YOUR EDGE

One common feature of all successful trading methodologies is that they invariably provide the trader with an edge. In the pit I had many edges, most having to do with my physical proximity to the order flow. Because I stood so close to the people executing the orders that I could smell them, I had first shot at the most lucrative opportunities and first shot at getting out when I was wrong. In an electronic environment, one needs to find other edges: making sure that you have the fastest connections to the exchanges, the best information flow, profitable trading strategies, and reasonable risk management techniques. The best way to understand the concept of an edge, ironically, is to think about trading in terms of gambling. Consider a wheel of fortune with 100 slots: 49 red, 49 black, and 2 slots reserved for the house. In this game every time the wheel is spun the gambler has a 49 percent chance to win against the house's 51 percent chance. While the gambler may get lucky and win for a while, if she plays long enough, she will eventually lose everything to the house. I am not suggesting that one must quantify the percentage chance of success prior to making a particular trade (although certainly systems traders are able to perform such calculations). What is important, however, is to develop an approach that gives you at least a fighting chance of beating the house. Every good trader has an edge. How that edge is cultivated is the ultimate determinant of how successful one becomes.

"Always Defer to the Lighthouse"

There's a great story that made the rounds a while ago on the Internet (I received it from at least a half dozen people who spend their time circulating e-mail wisdom) and while it has nothing to do with trading per se, it is a perfect metaphor for the problem that inhibits many traders from reaching their potential. It seems that two battleships were on patrol in heavy seas. The captain who tells the story says that he was serving on the lead battleship and was on watch as night came. Visibility was poor, so he stayed on the bridge watching for any potential problems. Suddenly a lookout informed him that there was a light on the starboard bow. The captain asked, "Is it steady or moving astern?", to which the lookout replied, "Steady, Captain." The captain was concerned because this meant that the two ships were on a collision course. He called to his signalman, "Signal that ship: We are on a collision course, and advise you change your course 20 degrees." A signal came back: "Advisable for you to change your course 20 degrees." The captain said, "Send the following message: 'I'm a captain, change your course 20 degrees." The reply came back: "I'm a seaman second class, you had better change your course 20 degrees." The captain was furious by that point and said: "Send, I'm a battleship. Change your course immediately." The response that was flashed back was as follows: "I'm a lighthouse." The captain changed course. Whether this event actually happened is immaterial, as we can learn a good lesson from it. The best traders are those who see a situation and understand all the dimensions of it. When one assumes he or she knows everything, or worse, refuses, out of arrogance, to consider the ramifications of a decision based on incomplete information, the end result is seldom a good one. In fact, in some percentage of cases, the result will be disastrous. If you want to trade successfully, you must develop a sense of perspective. As the captain understood, always defer to a lighthouse.

SYSTEMS-BASED TRADING VERSUS DISCRETIONARY TRADING

There is a certain appeal to developing systems that require you to act in a specific way every time market movement sets off a signal. In order to derive the full benefits of a profitable system, the trader must carry out the instructions to the letter. For some traders it is far easier to maintain a discipline about their trading when there is no discretion in the decision-making process.

Developing systems, however, is not without its pitfalls. While some traders are able to use programs such as *Trade Station* or *Meta Stock* to back test historical data, many run into problems utilizing the software. One impediment is that everyone seems to be searching for the holy grail of systems; a system so powerful it is 100 percent foolproof and so secret that it has not been discovered. Someone is likely to find the holy grail before he or she finds such a system. Another problem is that, very often, the market data that is available is corrupted with inaccuracies. A third problem, probably the most significant of all, is that many developers tend to *optimize* their systems to the point where they are of no predictive value (Optimization occurs when the developer knowingly or unknowingly "fits" the market data to support a particular conclusion).

Discretionary traders have rules to follow, just like their programming counterparts, but apply those rules based on their subjective understanding of the situation. From trade to trade there may be significant differences in how the trader's strategy is implemented. This is not to imply that the discretionary trader can get away with lapses in discipline. In fact, because his decision-making process is subjective, discipline must be in the forefront of his mind at all times.

So the question arises, which approach is best: systems-based trading or discretionary? The answer, which you knew of course, is that there is plenty of room in the trading world for both types of traders. The important thing is to think carefully about which approach will better suit your needs and then take the necessary steps to become proficient at it.

UNDERSTAND THE VALUE OF CAPTURING THE "SPREAD"

What most active traders understand, and many people ignore, is the value of capturing the spread; that is, buying at the bid price and selling at the offer price. In the pit environment, taking advantage of the spread is the trader's greatest edge and is largely unavailable to the customer. In electronic trading, by contrast, anyone with direct access can buy on the bid or sell on the offer with equal ease. What this means is that countless millions of dollars that would otherwise go directly into the pockets of the floor-trading community are up for grabs in the electronic environment.

Consider how important an advantage capturing the spread can be. Let's say a trader makes 100 round-turn trades per day for a total of 200 transactions (100 buys and 100 sells). If the average spread between the bid and ask is $25 and she never captures the spread, she will pay $5,000 ($25 × 200 transactions) every single day, or $1,250,000 per year. You may think that someone who trades less frequently need not worry about paying away the spread. But consider this: a person who pays away a $25 spread on 10 contracts per day will forego $125,000 every year.

While I am not suggesting that one can capture the spread on every transaction, it should be viewed as a direct cost of doing business.

Some business costs are fixed and you have no control over what you pay. The spread is different. To the extent that the trader is willing to develop an understanding of when it is appropriate to try to capture the spread, he can significantly lessen his transaction costs. Moreover, as the number of transactions grow, it is clear that the difference between capturing the spread or paying it away can be the difference between success or failure.

Some traders will argue that they cannot be bothered with capturing the spread. They are concerned that they may miss a market opportunity. What good is trying to save a $25 spread if you miss the chance to buy at a particular price and then have to pay $200 or $500 more to chase the market? Similarly, traders looking for a large move may say that, if their profit objective is $10,000 from the entry point, paying $50 to enter and exit the trade is a reasonable percentage to pay in order to ensure an execution. Even active traders who make their living off of capturing the spread oftentimes will hit the bid or buy the offer if the situation warrants quick action. While these are all reasonable points, the important thing to recognize is that, whenever possible, you should use the technology available to capture the spread because doing so can significantly improve your profitability.

CONTROL YOUR RISK

The most important element of the trading strategy, by far, is risk management. If more traders understood this, there would be more successful traders. I speak to a lot of new traders and I will often test them with the following question: Which is the better bet; you have a dollar and can bet it all to make two dollars, or you can bet a penny to make two pennies? As stupid as most new traders are, they are not stupid enough to get such a leading question wrong. Thus far, no one has ever tried to tell me that they'd rather risk blowing their whole stake in an all-or-none bet. Yet, in practice, too many traders—veterans, as well as novices—repeatedly put on unjustifiable positions, which, in an adverse market move, could wipe out all their capital. Perhaps they think big bets are the quickest way to riches. Rather, they are the quickest way to a margin call.

While I have always understood this intellectually, my resolve has been tested many times. One occasion, in particular, sticks out in my mind. Shortly after I made my move onto the top step of the Deutsche mark pit—and it seemed awfully important to me to be seen as a big trader—the order filler for a large money manager walked into the pit and asked me to make a market. This broker's orders were coveted because he was known for not haggling about the price, and here he was offering to trade with me ahead of everyone else in the pit. All I had to do was make a market, so I told him the bid was 17 and the offer 18. He promptly sold me 1,000 contracts at 17 and walked out of the pit before I could say another word. At the time, I had $50,000 in my trading account and had just bought $75 million worth of Deutsche marks—20 times my normal position size. You do not need an advanced degree in economics to understand the colossal recklessness and stupidity of what I had done. I did have the presence of mind to know I needed to get out of the position immediately and that I needed to do it as quietly as possible. If the pit knew that I needed to get out of a trade that large, they would squeeze the life out of me. As it happens, the broker for Goldman Sachs and the broker for Morgan Stanley were each bidding 17 on 300 contracts. In the split second that I thought about my options I realized that I should sell the bids to scratch 600 contracts at 17, but wondered what I was going to do with the remaining 400. Ironically, by hitting the two bidders and loading them up with $45 million of Deutsche marks, I would likely end up competing against them to get out of the rest of my position. If that were the case, the 600 contracts I scratched could end up being the weapon of my destruction. Luckily, at that moment, the Bear Stearns broker, seeing that I had traded with his rivals, bid 17 on 500, not because he wanted them, but to show me how upset he was that I had left him out of what must have been a good trade. He never knew what hit him when I said SOLD! And he never knew that I vowed to donate to charity whatever I was about to make on the additional 100 lots I was now short. Years later, as I tell this story, I still get the shakes. I put my career on the line out of hubris; to impress an order-filler with my market-making prowess. Had the market moved four *ticks* lower—which can happen in less time than it takes to sneeze—my trading account would have been wiped out. I vowed from that day forward that I would never put myself in such a vulnerable position again.

   Another time, years later, an order-filler came into the pit on a Friday afternoon five minutes before the close and tried to buy 1,100 contracts in the fourth expiration month of the Canadian Dollar. At that time, the entire open interest in that particular contract month was less than 500. Understandably, no one wanted to make him a market and he was forced to bid the spread, between the front month and the expiration he needed, 60 points beyond where it theoretically should be trading. The pit, which often clears out early on Friday afternoons, was virtually empty and I could have had the entire trade if I wanted it. I did the math quickly in my head and realized that this trade was worth more than $600,000 . . . theoretically. If the movement in the spread was an anomaly—simply the result of a huge order coming into an illiquid market—this trade was a slam dunk. But I asked myself, What does this guy know that I don't? Are interest rates changing in Canada? Is there trouble in Quebec? It took only a split second, but I considered the worst case scenario and knew that I did not want to be on the wrong side of this trade. So, instead, I did only 100 contracts; an amount that was manageable for me. Somehow the order-filler found others who were willing to take the rest of the trade before the market closed for the weekend. The trade was, in fact, a slam dunk and everyone who participated in it appreciated a windfall profit when they got out on Monday morning. When I've told this story to fellow traders, very few of them think I did the right thing. They believe that I should have taken the entire trade and that my fear of the trade size was irrational. I, on the other hand, know that I made the correct decision. After twenty years of trading, I have seen far too many "sure things" turn into re-creations of the voyage of the Titanic. I am determined to make it through my career without having such life-changing experiences, even at the expense of having to pass up the occasional windfall.

Among sophisticated trading houses and money managers, there are some very complicated risk management models. I favor a much simpler approach: Never establish a position that could cost more than 2.5 percent of your account equity. With that guideline in place, it would take 40 consecutive losing trades of the maximum amount to wipe out all of my capital (In practice, I have never lost 2.5 percent on a single trade, although I was always prepared for that outcome). It is almost inconceivable that I could suffer so many losses in a row. At the very least, it is likely I would abandon my system in favor of another approach—gazing at the moon, perhaps?—long before the fortieth loss.

Please note that the 2.5 percent threshold is nothing more than an arbitrary guideline with which I am comfortable. It is perfectly reasonable to budget no more than 1 percent per individual trade. Using that threshold, one could make 100 losing trades in a row before busting out. It is also reasonable, depending on the trader and trading style, to risk more than 2.5 percent on single trades. I know traders who are willing to risk from 5 to 10 percent. While this is an extremely aggressive approach, for longer term trades or in volatile markets, such extremes may be appropriate. One caveat, if you are trading multiple markets, is that you need to be honest about the true nature of your position. A long position in two markets that are substantially correlated, such as Bonds and 10-Year Notes or Soybeans and Soybean Oil should, for risk purposes, be viewed as a single position.

UNDERSTAND WHAT YOU ARE GETTING YOURSELF INTO

Trading is serious business and the risks are real. There are no "do-overs" and no sympathy if you screw up. If you are already trading you are aware of this, and if you are opening an account soon, you will read about it in the disclosure document you will receive from your FCM. At the risk of scaring you away from the market entirely, understand that when you trade with futures leverage you can lose everything you have in your account and then some. You need to understand the worst case scenario and be willing to live up to your responsibilities if you are not successful.

I am constantly amazed by the number of traders who, when they lose big money—life-changing money—look for someone else to blame for their troubles. Sometimes there is a culprit—a negligent broker or dishonest counterparty—and in those cases, I hope that you can find the evildoer and force him, in the appropriate disciplinary forum, to take responsibility for his actions. But more often than not, the reason for a loss is far less insidious. You messed up. You took on too many contracts or stayed with a position too long. You got greedy, stupid, or complacent, or maybe all of the above. The question is, if and when such a disaster occurs, how are you going to deal with it? Because we live in a litigious society, the way losers often deal with it is to call a lawyer and sue someone—usually whomever has the deepest pockets. This should turn the stomach of every responsible trader; not only because it is wrong, but also because it raises the cost of doing business for everyone in the trading community. Every dollar spent by an FCM or IB in fighting a frivolous lawsuit is a dollar in cost that is ultimately passed along to others in the trading community.

Probably the most extreme instance of this phenomenon is being played out right now. In January of 2001, a jury, by a vote of 10-0, awarded $164.5 million to a former Bear Stearns customer named Henryk de Kwiatkowski, finding that the FCM was negligent in its handling of de Kwiatkowski's *non-discretionary* currency trading account. The case began in 1991, when the investor opened an account to trade currency futures. At that time, he said his net worth was in excess of $100 million. During 1991 and 1994, de Kwiatkowski was quite the active futures trader, netting more than $447 million in profits. But, between December 1994 and March of 1995, he lost all of that and an additional $7 million for good measure.

Here is how it happened: In October of 1994, de Kwiatkowski instructed Bear Stearns to sell 65,000 currency futures contracts at the Chicago Mercantile Exchange. The notional value of these contracts was in excess of $6.5 billion. In late November, the head of Bear Stearns currency trading unit recommended to de Kwiatkowski that he transfer his trades to the cash market, which was better suited to such a large position. He partially complied and switched half of his position from the CME during the first week of December. By that time, de Kwiatkowski must have felt like Christmas had arrived early, as his positions had appreciated in value by $228 million. But in late December, the dollar fell sharply. On December 26 alone, he lost $110 million. Later, on January 9, he lost $98 million; and on January 19, another $70 million. At that point, had he liquidated his positions, de Kwiatkowski would have been left with $34 million in profits. Stubbornly, he refused to get out, and a little more than a month later, when additional losses piled up, Bear Stearns liquidated 18,000 positions to satisfy a margin call. de Kwiatkowski later approved that liquidation, but nothing more. The next trading session brought more losses, and finally, after $226 million in trading losses in six weeks, de Kwiatkowski authorized the liquidation of his position.

de Kwiatkowski filed suit against Bear Stearns in 1996, stating that his original position in the market was based on a bullish dollar recommendation made by Wayne Angell, one of the firm's chief economists (and a former Federal Reserve governor). Later, in February, two other Bear Stearns economists issued a bearish recommendation, which the firm mailed to de Kwiatkowski's home in the Bahamas. But de Kwiatkowski said that he never saw the report and that his butler often did not give him all his mail in a timely way. Therefore, he alleged in the suit that Bear Stearns had breached its fiduciary responsibility to him by not making him aware of the new report. Bear Stearns argued that, because de Kwiatkowski managed his own account and was a sophisticated investor, they were not obligated to cover his speculative losses simply because they offered some ancillary investment advice gratis. When the jury awarded de Kwiatkowski his victory, Bear Stearns could not believe it, and planned to take the verdict before an appeals court.

THE INVESTOR'S BILL OF RIGHTS

Certainly there are legitimate instances of broker fraud and negligence that require redress. In order to help customers understand their rights, the National Futures Association created the *Investor's Bill of Rights*. This document spells out in unambiguous terms what the customer can reasonably expect from his or her broker. (For the complete text, see Appendix E.) The key provisions are:

- *Honesty in advertising.* The only bona fide purposes of advertising are to call attention to an offering and encourage you to obtain additional information.
- *Full and accurate information.* All material facts about an investment need to be disclosed. Also, the investor is entitled to request information about the principals of the firm with which the account is being opened.
- *Disclosure of risks.* Every investment entails a certain amount of risk. You are entitled to understand the risks of an investment prior to investing.
- *Explanation of obligations and costs.* You have the right to know all the obligations and costs associated with an investment. You should not be hit with additional charges of which you were unaware at the time of the investment.

- *Time to consider.* You should never be pressured into making an investment. High pressure sales tactics are inconsistent with a professional approach to customer solicitation.
- *Responsible advice.* If an investment professional proffers advice, you have the right to expect that it be responsible. This means that the broker must "know" the customer and only sell him investment products that are suitable for him.
- *Best effort management.* If you give a firm money to manage, you are entitled to have the money managed responsibly. The manager may not make unauthorized investments, excessive numbers of investments for the purpose of creating additional commission income, or appropriate your monies for personal use.
- *Complete and truthful accounting.* It is your right to obtain a complete accounting of your account activity on a regular and timely basis.
- *Access to your own funds.* You have the right to be informed of any restrictions with respect to obtaining immediate access to the money in your account. You are also entitled to have access to the individual in the firm who controls the disbursement of funds.
- *Recourse if necessary.* You are entitled to seek an appropriate remedy if someone has treated you dishonestly or unfairly. You should understand when you open your account what remedies are available to you in the case of a misunderstanding or alleged wrongdoing.

LOSSES: CAN'T LIVE WITH 'EM, CAN'T LIVE WITHOUT 'EM

Because trading is a zero-sum game, there is a loser for every winner. With millions of transactions every day chances are, from time to time, that you will be on the losing end of some trades. When you first start you may be on the losing end of most of your trades. Do not get overly concerned about this. It is an unavoidable part of the process and every trader must find a way to deal effectively with losing money. There are a couple of aspects to this. First, of course, is dealing with the practical concern of, "Oh, God, I've lost money, how am I going to pay my bills?" The second is dealing with the emotional component of, "Oh, God, I've lost money, I am the biggest idiot in the history of trading;" or "Oh God, I've lost money, I can't tell my spouse." These are real concerns, to be sure. Nevertheless, you cannot allow things over which you have no control—for example, money that is already lost—to influence your subsequent actions. To conserve capital, good traders manage their risk responsibly. As for the emotional component, they learn that losing money does not, by itself, make one an idiot. Not telling one's spouse, however . . .

This advice is often easier to give than to take. When I was a young trader, I did not have much composure when it came to taking losses. In retrospect, I think the problem was that after a couple of months of losing, I began to equate my self-worth with my net-worth, a very common problem among new and established traders alike. I was unhappy at work and even unhappier outside of work, where I stewed at night over my incompetence during the day. I remember having a recurring dream throughout that period: The market opens and somehow I know that I am long 100 contracts. This is absurd, of course; not only am I too new to take such a large position, but the unemployment figures are to be made public in five minutes and I am always market-neutral going into an economic release. Yet, unbelievably, the market is going in my direction. It rallies 10, 20, 50 points! I am making money for the very first time, and we are talking serious money. I decide to get out and take my profit, so I turn to the trader next to me and say, "I'll sell you 100." There is no response. I say it a little louder and he continues to ignore me. I am angry now and scream at him, "I'll sell you 100!" Still nothing. By now, I am getting nervous. I go to another trader and try to get out. Again, no response, and after trying a few more traders I realize that no one is going to buy this 100 lot from me. At that moment, the figures are released and they are bearish beyond expectation. As the market plummets and my windfall profit turns into a loss of epic proportion, I grab a trader by the knees and beg him to take the trade. It is then that I hear it; they are all laughing at me while the market continues lower, lower, and lower still. At that point, I wake up, sweating and hyperventilating, ready for another exciting day in the pits.

What I did not understand, and came to realize later, is that the bottom line is the last place one should look for personal validation. It is perhaps one of the hardest things to accept; no trader is perfect and taking losses is part of the challenge of the job. In fact, if one is truly enlightened it is possible to see a benefit in losing. *Why are you banging your head against the wall?—Because it feels so good when I stop,* goes the old joke. So too with losing. I appreciate every loss because it feels so good when I stop. Looking back at my career, I can say that some of the best trading days I ever had are ones which ended in the red. I recall once buying a large number of Japanese Yen contracts during a fast market. Before I could record the transaction, the next trade was 10 points lower. I immediately hit the bid to take my loss. It was a good thing I did, as the next trade after that was 50 points lower thanks to an unexpected intervention by the Bank of Japan. During the rest of the day I traded actively, but not because I was trying to make back the loss all at once. I traded because it was busy and whether I was up or down was immaterial; a trader trades and tallies the results only when there are no more trades to be made. I went home a loser that night—I actually lost quite a lot of money—but slept perfectly well; no psychotic dreams or night sweats. If you think that the source of my quietude was the fact that I managed to avoid a cataclysmic loss of 60 points per contract, you are missing the point. Had I not been successful in hitting that first bid 10 points lower, I would have lost much more money, but nothing about how I reacted in the aftermath of the loss would have changed.

As unlikely as it sounds, if you can simply learn to conquer the very natural feelings of self-loathing that arise out of repeated failure, you will be well on your way to becoming a successful trader.

TECHNICAL ANALYSIS

There are many excellent books that one can buy to study technical analysis (see the bibliography in Appendix G), complete with detailed explanations about everything from mundane trendlines to the most esoteric charting techniques the human mind can conceive. This is not one of those books. Still, technical analysis, has been integral to my success as a trader and it would be difficult to write a book about trading without even mentioning the subject. A great debate has raged, seemingly forever, about whether charting works, and it is on that topic that I'd like to try to shed some light. I have spent much of my career trying to figure out whether I believe in technical analysis, or if all of us who rely on it are not just in a huge state of denial. Can a bunch of lines that represent past market performance really help us determine what is going to happen next? Most of the time I think they can. Sometimes, however, I think I would be better off calling the Psychic Hotline, where for only $1.99 a minute—far less than my market data costs—I can find out whether I am going to meet that special someone and where the Yen is heading. Usually, my doubts are sparked by some academic article purporting that the market is a *random walk*—completely unpredictable. But it hardly takes heavy quantitative research to shake my convictions. This excerpt from a recent column about investing, by humorist Dave Barry, is enough to do the trick[1]:

TOM BROKAW: The stock market today went either down or up and nobody on this Earth knows why. For more, here's our financial expert.

FINANCIAL EXPERT: Tom, analysts attributed the movement of the market to a market movement, in which the market moves either upward or downward, depending on the direction of the market, although sometimes it holds still.

BROKAW: And this is expected to continue?

EXPERT: Tom, it's too soon to tell.

What would I say to Dave Barry about technical analysis? Could I convince him that there is a logical foundation to the research chartists perform before making a trading decision? I suspect not, because even after twenty years of chart-watching, I have yet to entirely convince myself.

THESE BOOTS ARE MADE FOR RANDOM WALKING

One of the first, and best, books I ever read about the financial markets is *A Random Walk Down Wall Street* by Burton G. Malkiel, a professor of economics at Princeton University. This book is a classic and has been reprinted six times since its initial publication in 1973 (most recently in 1999). It seems that current generations and many more to come will be able to enjoy Professor Malkiel's categorical rejection of

---
[1] Dave Barry, *"How to Stay in the Black,"* Chicago Tribune Sunday Magazine (May 6, 2001).

the notion that past market performance can help predict where prices will head in the future. In answering the question "What is a Random Walk?" at the very beginning of his book, Malkiel states his position clearly:

> A random walk is one in which future steps or predictions cannot be predicted on the basis of past actions. . . . Investment advisory services, earnings predictions, and complicated chart patterns are useless. On Wall Street, the term "random walk" is an obscenity. It is an epithet coined by the academic community and hurled insultingly at the professional soothsayers. Taken to its logical extreme, it means a blindfolded monkey throwing darts at a newspaper's financial pages could select a portfolio that would do just as well as one carefully selected by the experts.[2]

Malkiel's vitriol against technicians in pin-striped suits, whom he compares to "bare-assed apes," is fueled by his deeply held belief that price history contains no useful information that enables an investor to consistently outperform a buy and hold strategy. While he acknowledges that trades entered into as a result of technical analysis do occasionally make money, he argues that such feats hardly prove that there is value to chart-watching. If, for instance, one administered a drug to a group of sick people and a placebo to another and found that both groups got better, it would be illogical to conclude that the drug was the reason for the improvement, even though the people who took it were no longer sick. The same holds true with technical analysis: the placebo, he argues, is like the buy and hold strategy, which produces substantially the same returns as trades based on trendlines, moving averages, and other technical schemes. Malkiel says that unless someone can prove technical schemes produce better returns than the market, they must be considered ineffective.

Although I use technical analysis every day and rarely make a trading decision without at least taking a peek at a chart or two, whenever I reread Malkiel, an alarm goes off in my brain; it is the one you hear whenever you leave on vacation and are pretty sure you turned off the sprinkler. You know things are probably okay, but as you drive onto the Interstate there is that nagging doubt that you might have made a colossal mistake. Malkiel presents four logical reasons that charting cannot

---

[2] Burton G. Malkiel, *A Random Walk Down Wall Street* (W.W. Norton & Company, 1998), p. 24.

possibly work. First, the chartist acts only *after* a price trend has been established. Because markets often move quickly, the chartist will miss the bulk of signaled moves. Second, because charts are available to everyone in the market, their predictive value, to the extent it exists at all, is diminished. Presumably, a signal that everyone acts on simultaneously is ultimately self-defeating. Third, he correctly points out that traders tend to anticipate signals to avoid entering the market at the same time as the other chart-watchers. For instance, if a buy-order is indicated at a price of 10, a trader may try to beat the competition by placing his order at 12. This leads other traders to place their buy-orders at 13 and still others to step out and buy at 15. These almost-comical attempts to outsmart the competition—to out-anticipate the anticipators—can theoretically go on until infinity. In practical terms, however, they only continue to the point where any advantage from the chart signal is undermined by the premium the trader pays to be first. Finally, Malkiel tells us that in the battle to be first an insider will always beat out the chartist; as he says, "If some people know the market will go to 40 tomorrow, it will go to 40 today."

As if these reasons are not compelling enough to abandon my charts, in doing the research for this book, I performed an experiment that would warm the cockles of a Random Walk theorist's heart. (I learned of this experiment as a young trader, from Arthur Sklarew's *Techniques of a Professional Chart Analyst,* Commodity Research Bureau, 1980—one of the first books I ever read about technical analysis. The book still holds up well more than twenty years later.) I prepared 102 small paper squares, numbered consecutively from zero to plus 50 and from zero to minus 50. I mixed the pieces up in a baseball cap and drew out a number, which I recorded (-7). I then placed the numbered square back into the box, shook it again, and drew a second number (-11). I continued picking numbers in this fashion and recording them until I had done so 250 times, which is about the same number of trading days as in a year. In order to convert the list of net changes to a list of simulated daily closing prices, a nominal starting price of "1,000" was chosen and the plus or minus figures were added or subtracted in succession.

We observe something very interesting from this exercise in Exhibit 9.1. Even though the "closing prices" are clearly random, patterns emerge on the chart that would cause a technician, trading only at breakout and reversal points, to erupt into the following trading frenzy:

Day 70	BUY 1	825
Day 74	BUY 1	911

Position: Long 2 contracts @ an average price of 868.

Day 128	SELL 3	994

Position: Sold 2 contracts @ 994 for a profit of 252 points; Short 1 @994.

Day 133	SELL 1	995

Position: Short 2 @ an average price of 994.5.

Day 162	BUY 2	937

Position: Flat. Bought 2 @ 937 for a profit of 115 points.

Day 184	SELL 1	845

Position: Short 1 @ at 845.

Day 194	BUY 1	971

Position: Bought 1 @ 971 for a loss of 126 points.

Day 202	SELL 1	880

Position: Short 1 @ 880.

Day 204	SELL 1	876

Position: Short 2 @ 878.

Day 250	BUY 2	804

Position: Bought 2 for a profit of 158 points.

So, using some basic technical approaches, the technician is able to take 399 points in profit out of the market (see Exhibit 9-1). The only problem is that I picked the market out of a Chicago Cubs hat. Random market theorists would point at this and argue that prices in the markets are just as random as the prices generated by the experiment. They would conclude that one could no more make meaningful predictions about the market any more than one could determine the next slip of paper to be picked out of a baseball cap.

To pile on the technicians further, if we go back to Malkiel, he tries to prove his argument by debunking the concept of the "hot

Exhibit 9-1 Random Walk Chart

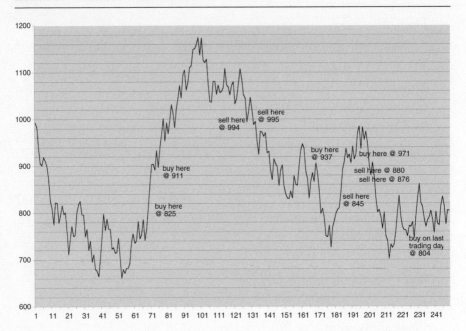

hand" in sports. Most people think, and athletes will insist, if a player has made his last shot or several shots in a row, that he should get the ball again to capitalize on his momentum. They believe, similarly, that a market moving in a particular direction—a market with a "hot hand," so to speak—has its own kind of momentum which is likely to continue until some material factor changes the momentum in the other direction. Malkiel, however, cites the work of a group of psychologists who demonstrated that this widely-held belief is a fallacy. They performed a detailed study of every shot taken by the Philadelphia 76ers over a season and a half and found no evidence of a positive correlation between one shot and the next. In fact, they found that a player was more likely to sink a basket *after missing a shot than he was to hit two in a row.* They concluded that while player perceptions were influenced by whether their "hands were hot," in reality, the outcome of the next shot is no more predictable based on the outcome of the previous one than the price of today's soybean futures are based on yesterday's close.

Malkiel goes on, constructing one argument after another to convince the reader that watching and trading off of charts is a foolhardy and futile effort. While there is an undeniable persuasiveness to his logic, he has not quite moved me to uninstall my technical analysis software and cancel my market data feed. It is possible that the market is a random walk, as he says, but that does not necessarily mean that there is no value to watching the charts and acting on the signals they produce. Random walk proponents focus on the inability of the technician to explain why technical analysis works. I admit that it would make me feel better about my trading decisions if I understood the how and why of them. Nonetheless, when I see moving average lines cross I know that a certain outcome can generally be expected. Although I have no idea why this is so, it does not stop me from using the information when it presents itself. The first time a human being stuck a hand into fire and screamed out in pain was the last time he did so. Certainly he had no understanding of the science behind fire, bare skin, and pain; but he learned his lesson and incorporated the knowledge into his everyday life and activities. It mattered little that he did not know everything. He knew what he needed to know.

I think it is possible to go farther in trying to refute the random walk concept, although I will still fall short of providing the type of explanation an economist, mathematician, or scientist would accept as objective proof. Consider this: There are thousands of traders with billions of dollars available to them for trading purposes. When leverage is factored in, particularly in the futures markets, the amount of money available for making trades is many times that. That money needs to be put to work trading. Charts offer traders a convenient, cost-efficient tool to help them organize their thoughts about what has happened in the market's recent past and what to do next. Malkiel contends that chart signals are pointless because the pricing information is so ubiquitous that the entire trading community is able to see the same picture at once. I believe that he has it wrong. The signals are valuable *because* they are ubiquitous. In other words, because the entire trading community can see what is going on, when there is a signal they will act. The opportunity to profit from a charting signal arises out of the trading activity borne of those actions. Were no one looking at the charts, there would be far less activity and perhaps no liquidity.

Let me elaborate by telling you one way I used to use a particular kind of chart. For many years in the Deutsche mark pit, I kept a point-and-figure chart. (It is not necessary to be in a pit to derive the same benefit I am about to discuss. The same principles apply sitting in front of a PC off of the trading floor.) Point-and-figure charts measure price only and disregard time. The charts are constructed by plotting prices in a series of alternating Xs and Os whenever the price changes by a specified amount. Decisions to buy and sell are based on a subjective understanding of how the movement of those Xs and Os form areas of support and resistance. In a normal trading day, I might form impressions from reading these endless lines of Xs and Os that would lead to my making tens of transactions. On a more volatile day, I might see signals that would lead to hundreds of trades. Were all of my interpretations valid or the resulting trading decisions correct? Of course not. I made many losing trades during that period. The importance of the charts, however, is that they helped me organize my thoughts and place the market's movement in a context that remained consistent over time. By utilizing the signals, which admittedly were subjective, I was able to place myself in the market and put my trading capital to work. I knew that if I controlled my losses and let my winning trades run, that simply by being in the market I would make money. Had I not been watching the movement on my point-and-figure charts I would certainly have made fewer trades and far less money. The Random Walkers argue that I would be better off by pursuing a "buy and hold" strategy. Perhaps, but that argument is not really relevant. I believe I will do better trading this way, and the growing equity in my account should be all the proof that anyone needs to appreciate this point. To take Malkiel's pill analogy and turn it on its head, just because a buy and hold strategy may be profitable does not mean that it is the only way to be profitable.

The other element that should be considered in trying to better appreciate the value of using technical analysis is market psychology. Notwithstanding the Philadelphia 76ers inability to hit two shots in a row (had the experiment been performed on Michael Jordan's Bulls, the results might have been different), I believe that there is momentum in the marketplace. When the market approaches an important support or resistance point, a savvy trader can almost feel the tension building. Consider the following scenario: If the market pushes through resistance at 17 and there is a stop order to buy 1,000 contracts at 18, anyone who bought the market prior to the stop being executed will have an opportunity to sell at a profit. On a superficial level, this is purely an issue of supply and demand. There is demand to buy at 18 and little or no supply; therefore, the market will continue upward until the demand is satisfied at a higher price. But anyone who has traded knows that how the market participants respond to the imbalance between supply and demand is what really determines how high that 1,000 lot order will take the market. The psychology of the traders in the marketplace, characterized by greed, fear, and uncertainty, tends to create momentum. To the extent that one can identify these momentum areas and act in accordance with the technical signals they produce, it becomes possible to be swept along in its path. Once again, it does not matter whether one is in the pit or at a PC; it is possible to feel these momentum areas building and exploding. It is definitely a skill—and not everyone will develop it—but for the sophisticated chartist, when the market starts to move through support or resistance, the feeling is palpable.

It turns out that I need not face the Random Walkers by myself. Robert Edwards and John Magee, in their classic text *Technical Analysis of Stock Trends,* first published in 1948, specifically address the issue of whether chart-watching is valid. They argue—as if anticipating 25 years before the fact that the Random Walkers would try to trample the ground they so lovingly cultivated—that "no one of experience doubts that *prices move in trends* and trends tend to continue until something happens to change the supply and demand balance[3]."

If you are able to view chart patterns and see within each price *tick* another trader's eyes staring intently at a screen, feel her quickened pulse when she thinks about making a trade, or experience the pressure of her sweaty hand on the mouse button as it clicks twice and turns a thought into an action, you will be able to use technical analysis as a tool to make money. Perhaps there is no science to it. Perhaps we are all just picking each other's pockets in a never-ending Ponzi game. Even so, if you can figure out a way to beat the other players, I suggest

---

[3] Robert Edwards and John Magee, *Technical Analysis of Stock Trends* (John Magee, Inc., 1992), p. 7.

you leave the debate to the academics and start looking for that next head and shoulders pattern. Jim Cramer, the money manager, CNBC commentator, and founder of the financial services *www.thestreet.com,* put it best in a commentary on his web site (*www.thestreet.com,* September 30, 1999):

> It wasn't me that turned the market into a casino . . . I can smell gambling when I see it and when we are out there taking Amazon up 6 points we are gambling and we should own up to that. We are assessing the odds and making a wager. I can't stand the sanctimonious people out there who would call this something else in a vain attempt to dignify what they do for a living. The worst are the ones who think there is a science behind it, a Nobel prize winning science. They are totally bogus. It is about an edge. It is about being early or ahead of the crowd. It is about perception and psychology and capitulation. It ain't about P/Es and Price to Book anymore . . . maybe it never really was.

10

TWENTY-FIVE FREQUENTLY ASKED QUESTIONS ABOUT ELECTRONIC TRADING

In the course of doing business with customers at my firm, Aspire Trading Company, my partner and I are asked questions every day about the ins and outs of electronic trading. Some of the questions relate to personal situations or arcane interests and are not necessarily relevant to the needs of the average trader (e.g., you cannot scalp e-minis in your 401K). But many of the questions relate directly to the most important topics reviewed in this book. As a way of summarizing the key issues, I have distilled the vast number of questions we received down to 25. While the list is by no means comprehensive, it is a good starting point for anyone who got lost along the way, as well as a helpful review for those who may have missed a point here or there. If the answers to these questions are unclear, you want more detailed information, or you have questions of your own, I invite you to contact me by phone at (312) 638-5191 or by e-mail (*david2740@aol.com*), and within 48 hours I will make sure that you have the information you have requested.

1. HOW MUCH MONEY DO I NEED IN ORDER TO TRADE ACTIVELY?

An active trader should have at least $50,000 available. It is possible to open an account with less, but I have found that traders who do not have at least $50,000 in their account worry about every *tick* so much they cannot concentrate on making smart trades. Furthermore, with

$50,000 available it becomes possible (when one is ready) to trade multiple contracts in multiple markets, thereby maximizing the potential for appreciating a higher return. You should also allocate an appropriate amount of money for the expenses you will incur in the first six months of trading. That amount will vary depending on the type of connectivity, computer equipment, and hardware you choose and whether you are trading in an arcade. For further information, see question # 18.

2. HOW LONG WILL IT TAKE ME TO BEGIN MAKING MONEY?

Everyone is different, but as a general rule, it will take an active electronic trader at least 3 to 6 months to be in a position to make money consistently. In some cases it takes longer, but certainly, within 12 months, a trader should be on the road to profitability.

3. HOW MUCH MONEY CAN I MAKE?

It depends on a number of variables (e.g., how much risk you are willing to assume, which markets you trade, and whether you are trading with direct access), but with $50,000 on account and trading full-time, one should aim to make at least $200,000 per year. This equates to about $1,000 per day. When one is more experienced and/or if one trades with a larger account, it is possible to make multiples of that amount. The best traders make millions of dollars every year, using many of the same strategies that the average trader uses to make substantially less. Please understand that many people who try to trade futures are not successful.

4. HOW MUCH MONEY CAN I LOSE?

Futures trading is very risky. You can lose everything in your account and then some. This is why it is important that you carefully read the risk disclosure forms provided by your broker and that you understand what the documents say before you make your first trade. Many traders, even those who have experience in the traditional Open Outcry markets, lose money in their first months of online futures trading. Nevertheless, with a proper risk management plan in place, a trader should be able to protect himself from the calamitous events that wipe out trading accounts.

5. ASSUMING ONE HAS TO PAY HIS OR HER DUES, HOW MUCH MONEY CAN I EXPECT TO LOSE IN THE FIRST FEW MONTHS OF ONLINE TRADING?

It is not unusual for new online traders to lose $10,000 to $20,000 in their first six months of trading. If you become successful, this will be the best investment you ever made in your life, and you will be paid back many times over.

6. HOW MANY CONTRACTS SHOULD I TRADE?

Before you begin trading you will discuss your trading limits with your broker. The simplest approach is to divide the margin amount for the instrument to be traded into the account size and derive a trading limit. For instance, if the margin on an e-mini is about $5,000 and one has $50,000 in an account, the limit of outright longs or shorts is 10 contracts. There are circumstances, however, in which the trader and broker may wish to adjust the limits either higher or lower. It is very important for the trader and broker to communicate on an ongoing basis to determine the appropriate amount of capital needed to trade the number of contracts necessary to meet the customer's needs. Whatever the agreed-upon limit, in an electronic environment the broker is able to program the amount into the trader's risk management profile so that if the programmed limit is 10 contracts and the trader tries to buy 11, the trading software will reject the order.

7. WHEN I BEGIN, HOW MANY CONTRACTS SHOULD I TRADE?

It is strongly recommended that, irrespective of the size of the trading account, traders begin by trading one-lot positions. So, for instance, in the e-mini S&P 500, you should trade one contract (which, at this writing, is the equivalent of about $65,000 worth of stock). It is important to learn how to use the keyboard and mouse and develop your self-confidence before stepping up your volume. When you are comfortable and confident enough to trade without thinking about the physical mechanics of making the trade, then you can raise your position size. For some, it may take days; for others, it can take weeks or months.

8. HOW MUCH SHOULD I TRY TO MAKE ON A TRADE?

It is strongly recommended that a trader look for trades with at least a 2:1 risk-reward ratio. This means that if you are willing to take a loss of 1 S&P point on a trade if you are wrong, then you should have a reasonable expectation of making 2 points if you are right. The 2:1 ratio is a good guideline for short-term trades. For longer-term trades, it is suggested that a 4:1 risk-reward ratio is probably more appropriate.

9. WHAT IS THE MAXIMUM LOSS I SHOULD TAKE ON A TRADE?

The simplest approach is to choose a percentage stop based on the position size and the amount of money in the trader's account. For example, if one has $50,000 on account and has a 2 percent stop loss limit, a position should not be allowed to lose more than $1,000. Therefore, if one wants to be trading long with the e-mini at a 2-point stop (each e-mini point is $50, so a 2-point stop is $100), the maximum position size for that trade would be 10 contracts. The percentage one is willing to lose per trade is a somewhat arbitrary determination. The smaller the percentage, the more losing trades one can make.

10. WHAT ARE THE BEST ELECTRONIC FUTURES MARKETS TO TRADE?

The following electronic futures markets offer the best trading opportunities:

- E-mini S&P 500 (CME)
- E-mini NASDAQ 100 (CME)
- Treasury Futures (CBOT)
- Currency Futures (CME)
- Long-Term Interest Rate and Stock Index Futures (EUREX)
- Short-Term Interest Rate and Stock Index Futures (LIFFE)
- Interest Rate and Stock Index Futures (SFE)

EUREX and LIFFE products are problematic for U.S. traders because of the time zone differences. The heart of the European trading day is the middle of the U.S. nighttime. However, Sydney is an interesting market for U.S. traders, because the Sydney trading morning commences at about the time the U.S. stock index markets begin to die down.

11. HOW MUCH SHOULD I PAY IN COMMISSION?

It will vary based on a number of considerations, including how much volume you produce, which markets you trade, and whether you are in a trading arcade or at a remote location. As a general rule, an online futures trader should expect to pay less than $10 per round-turn transaction, including exchange fees. For active traders, the round-turn commission is likely to drop below $5.

12. WHAT ARE MY CONNECTIVITY CHOICES?

There are three different types of connections:

- A direct connection to the exchanges
- Single-pipe service, which entails a direct connection to a broker who is connected directly to the exchanges
- Connection to a broker or an exchange through an Internet Service Provider

13. WHICH CONNECTION IS BEST?

It depends on your needs. For high-volume, high-turnover traders, a direct connection is highly recommended. For those who trade infrequently, an Internet solution is probably sufficient. Single-pipe service is not as fast and efficient as a direct connection, but it is faster, more secure, and more stable than trading over the Internet. Because direct connection entails substantial upfront costs, the best way to enjoy the benefits of it is to trade in an arcade setting. By splitting the costs among the arcade users, it is possible to appreciate significant cost savings.

14. WHAT IS THE FIRST STEP I HAVE TO TAKE IN ORDER TO GET CONNECTED?

The first thing to do is to find a broker who can service your needs. In the case of Aspire Trading, for example, we work with our customers to determine which connectivity solution best meets their needs. Then we help the customer get set up so that he or she can begin trading in the shortest possible time and with a minimal amount of worry and aggravation.

15. IF I DO DECIDE TO TRADE OVER THE INTERNET, WHICH SERVICE PROVIDER SHOULD I CHOOSE?

It is extremely important that you use a Tier I Internet Service Provider, such as Ameritech or UUNET. Tier I providers have very large, high-speed connections to the Internet, as well as multiple paths to the Internet backbone for redundancy. Do not use a provider such as AOL or Compuserve, which are simply private networks with gateways to the Internet.

16. CAN I CONNECT TO THE INTERNET THROUGH A 56K DIAL-UP LINE OR DO I NEED A HIGH-SPEED LINE?

It is highly recommended that you use a DSL line to connect to your ISP, if one is available. DSL lines are preferable to cable modems, because cable modems utilize "shared Internet bandwidth." This means that as many people use the bandwidth available to you, your allocation of it diminishes. A cable modem, however, is still better than a dial-up connection. There are two types of DSL: Asynchronous DSL (ADSL) and Synchronous DSL (SDSL). For a trader, SDSL is preferred.

17. WHAT TYPE OF PC DO I NEED?

Your PC should be at the upper end of what is available on the market at any given time. As a general rule, a trader's PC remains up to date for 9 to 12 months. It is important to upgrade your PC on a regular basis in order to keep up with the demands of new software releases and because the traders who are competing with you will be upgrading. Try to buy PCs that can be easily upgraded. In that way, you can keep the cost and aggravation of upgrading to a minimum. One strong recommendation is to buy a first-class flat-panel monitor, preferably with an 18-inch viewing area. It costs more than a conventional monitor (at this writing an 18-inch flat panel goes for about $1,000, while a 19-inch conventional monitor is about $400), but it is worth it. A flat-panel monitor consumes less energy, produces far less heat, and requires only a tiny space on the desktop. Most importantly, it is far easier on the eyes. Remember, your monitor is your window to the trading world. Pay a bit extra to make sure you get the best view possible.

18. HOW MUCH MONEY DO I NEED TO INVEST IN THE FIRST SIX MONTHS OF ONLINE TRADING?

A trader who wishes to trade in an arcade should expect between $1,000 to $1,500 per month of fixed costs, which includes:

- a space in the arcade
- a trading workstation
- the front-end trading system
- Internet access
- technical analysis package
- news
- squawk box services
- training
- technical support and maintenance
- access to the trading front-end over the Internet for times when the trader is not in the arcade

A trader who wishes to trade from a remote location through single-pipe service will have to provide his or her own PC (about $2,500 with monitor) and phone line (cost varies depending on the type of line. A T-1 connection can be many hundreds of dollars per month while a standard phone line goes for a nominal fee). The trader at the remote location will also pay a license fee to the broker for the front-end trading system (about $300 to $600 per month) and license fees for any other services on the broker's network to which he wants access (a technical analysis package can run between $100 to $400 per month). With an Internet connection, the remote trader needs to provide a PC and will also pay license fees for software similar to those of the single-pipe user, although the fees may be slightly lower. He will also have to pay for Internet service (about $25 per month) and a DSL or cable modem connection (about $50 per month).

19. DO I HAVE TO TRADE FULL-TIME IN ORDER TO BECOME SUCCESSFUL?

It is very difficult to succeed at anything when you only attend to it part-time. Trading is like any other endeavor: if you want to be good at it, you need to make a serious commitment. While occasionally you hear of a successful part-time trader, I am very skeptical whether anyone can become a successful trader without trading full-time. There are two things you should remember:

1. trading is a zero-sum game
2. your competition is comprised of individuals who have made a full-time commitment

20. HOW DO I CHOOSE A BROKER?

You should choose a broker that specializes in online trading, such as Aspire Trading Company. The key considerations should be:

- Does the firm understand the technology?
- Can they provide it to me at a reasonable cost?
- Will they help me become successful?

21. I HAVE TRADED SECURITIES ONLINE BEFORE, BUT AM NEW TO THE FUTURES MARKETS. WHAT ARE THE DIFFERENCES BETWEEN THEM AND WHAT MAKES FUTURES BETTER?

There are many inherent benefits to futures trading that make trading the instruments online attractive. These benefits are most readily apparent to those who trade the e-mini markets and include:

- Lower commissions
- Increased leverage
- Central Limit Order Book

- Level playing field for all market participants
- No prohibitions to, or restrictions on, short-selling
- Large daily ranges and "no decimalization"
- Preferential tax treatment and less cumbersome filing obligations

22. WHEN CAN I TRADE SINGLE STOCK FUTURES?

The official kick-off to Single Stock Futures (SSF) in the United States will commence on August 21, 2001 for professional traders and in December of 2001 for everyone else. The two entities that will make the biggest push to capture market share, at the outset, will be new for-profit online joint-ventures; one pairs the NASDAQ and LIFFE, while the other combines the CME, CBOE, and CBOT. It is questionable whether SSF will become viable products. While, theoretically, there is ample need for them, they are on dubious regulatory ground at the moment. Unless and until the CFTC and SEC, which jointly regulate the products, can agree on a number of difficult and divisive issues, it is hard to believe that SSF will attract the type of liquidity that will enable the average trader to have a reasonable chance of making money.

23. THERE ARE A NUMBER OF INTERNET-BASED FOREIGN EXCHANGE TRADING SITES. WHY SHOULD I TRADE CURRENCIES AT THE CME?

The currency products at the CME combine everything that an active online trader needs in order to make money. There is liquidity, generated by the members of the exchange and the Globex Foreign Exchange Facility (GFX); direct access to the market from Sunday afternoon until Friday afternoon (Chicago time); direct access to other CME products, such as e-minis, on the same screen; and the CME clearing house guarantees all trades.

24. HOW CAN I GET STARTED STUDYING TECHNICAL ANALYSIS?

The easiest way to get started is to read one of the many excellent books that have been published in the last few years on the subject (see Appendix G, Suggested Reading). If you live in Chicago, the CME and CBOT offer classes on charting techniques such as Candlesticks,

Market Profile, and Elliot Wave analysis, as well as more comprehensive classes that cover a wide range of technical topics. If you wish to develop trading models, there are many good books on the subject (again, see Appendix G) and many different software programs that allow you to test trading ideas without the need for extensive programming expertise. Among the best programs are those by Omega Research, Metastock, and Omni Trader.

25. WHAT IS THE BEST TECHNICAL STRATEGY FOR ME TO USE TO MAKE MONEY TRADING ONLINE?

There is no single strategy that is right for everyone. The best way to determine which technical approaches are appropriate for you is to learn about them all and start to trade the signals they generate. Among the technical strategies active traders often look for are traditional support and resistance patterns, such as "Head and Shoulders," "Triangles," and various types of "Gaps." Moving Averages, Relative Strength measures, and Volume and Open Interest studies are also popular. For those who are mathematically inclined, many sophisticated quantitative studies are included as part of the standard technical analysis package offered by the major vendors. In most cases, the software allows you to create your own mathematical studies as well.

APPENDIX A
Glossary

Agent An individual or entity that executes an order on behalf of a customer. For example, when a FOREX dealer is told by a hedge fund manager to buy $10 million worth of Japanese Yen, he executes the order as an agent of the manager.

Arbitrage The simultaneous purchase and sale of identical or similar instruments for the purpose of profiting from the price difference between the instruments. For example, if the Japanese Yen is offered on GLOBEX at 8500 and bid in the pit at 8505, one could purchase the offer in the electronic market and sell the bid in the pit for a guaranteed 5 point profit. Arbitrage opportunities tend to disappear quickly in all markets because, in trying to capitalize on the price discrepancy, market-makers drive the prices of the two instruments together: In our example, the buyers quickly buy the 8500 offer and bid the market; sellers hit the 8505 bid in the pit, driving the price in that market lower:

8500 8501 8502 8503 8504 **8505**
BUYERS $\Rightarrow \Rightarrow \Rightarrow$ $\Leftarrow \Leftarrow \Leftarrow$ SELLERS

Arcade A trading room in which traders gather together to share the costs of creating a professional trading environment. Also known as a dealing room.

Back office The business division of the FCM that is responsible for post-trade activities such as clearing and settlement.

Benchmarking How a money manager compares her performance to that of an index. Some common indices that are used for benchmarking purposes are the S&P 500, NASDAQ 100, Russell 2000, and Wilshire 5000.

Beta A measure of how sensitive a portfolio or security is to the movement of the entire market. For instance, a portfolio with a beta of 1.2 will move 1.2 percent for each 1 percent move in the market. Similarly, a security with a beta of .8 will move .8 percent for each 1 percent move in the market.

Bid The price that a buyer will pay for an instrument.

Broker-Dealer A firm authorized by the Securities and Exchange Commission to conduct trades in securities markets as an agent (executing orders on behalf of a customer) or a principal (for the proprietary trading account of the firm).

Buying power The amount of stock one can control based on the current total of capital on account with the broker-dealer. The Federal Reserve sets margin rates, and under its Regulation T requires stock to be margined at no more than a 2:1 ratio. Therefore, if one had $50,000 on account, that person would be allowed to purchase up to $100,000 worth of stock.

Capital Asset Pricing Model (CAPM) A mathematical way of explaining how securities should be valued, based on how risky they are compared to the return on a risk-free asset such as Treasury bills. From this model, we are able to calculate the beta of a portfolio or security.

Capitalization Weighted Index An index that is calculated using market weights; the share price multiplied by the number of outstanding shares. The S&P 500 Index is capitalization weighted.

Cash settlement Futures contracts that are settled with the transfer of cash rather than delivery of a physical commodity. Stock index futures are an example of a cash-settled contract. Soybeans are an example of a physically delivered contract.

Central Limit Order Book (CLOB) Any electronic system in which all orders are sent to, and matched in, the same location. The NASDAQ market does not utilize a CLOB; futures exchanges do. Systems with CLOBs tend to be more stable, efficient, and speedy than systems without.

Chicago Board of Trade (CBOT) One of the largest futures exchanges in the world. Its main products are grains and long-term interest rates. It has partnered with EUREX, the German futures exchange, to create an electronic platform known as a/c/e.

Chicago Mercantile Exchange (CME) The largest U.S. futures exchange. Its main products are short-term interest rates, stock indices, currencies, and meats. Its electronic platform is called GLOBEX.

Clearing The process of transmitting, reconciling, and confirming instructions prior to settlement.

Commodity Futures Trading Commission (CFTC) The governmental regulator of the futures industry.

Connectivity The method whereby one connects to the trade matching engine of a particular exchange's electronic system. The preferred method for active traders is by high-speed phone line, such as a T-1.

Convergence The tendency of a futures price to move toward the cash price of the underlying asset from which it is derived as expiration nears. At expiration, by definition, the futures and cash price will "converge" into the same price.

Credit risk The risk that market participants will default on an obligation to perform.

Day trader Typically a trader who looks for short-term opportunities in the market and rarely holds an overnight position.

Decimalization A securities pricing convention, implemented by the New York Stock Exchange and NASDAQ in 2001, which allows stocks to be traded in increments as small as a penny. This change to the time honored tradition of pricing stocks in fractions enables investors to achieve tighter bid/ask spreads than they did under the old system. It also helps non-professional customers to know the true price of a stock. While one may not understand how much 15 7/8 is, everyone can appreciate what $15.88 means.

Demutualization The act of converting from not-for-profit to for-profit exchange. The practical effect of the move is to de-emphasize the focus on the needs of the membership in favor of concentrating on building shareholder value.

Depth The number of outstanding orders pending on either side of the market.

Derivative An instrument that "derives" its value from an underlying asset. S&P futures, for example, are derivatives of the S&P Index.

Direct access The ability to participate on the current bid and offer with other market participants and to be able to see and have access to the order book and other useful market information.

Electronic Communication Network (ECN) An electronic trading platform that matches buy and sell orders that are routed to the platform. To the extent that an order sent to the ECN cannot be matched, it may be routed to a different trading venue (another ECN or stock exchange). Among the biggest ECNs are Instinet, Island, and Redi.

EUREX The German futures exchange, which is completely electronic. EUREX's main products include long-term European interest rates and European stock indices. EUREX has partnered with the CBOT in an electronic platform called a/c/e.

Fair value The concept of fair value arises out of the symbiotic relationship between the futures and underlying cash Index. The price between the two will always be different until expiration, when the prices of the two instruments converge. Traders attempt to determine what the fair value difference should be, based on the amount of time until convergence and prevailing interest rates. When traders perceive that the difference between the two instruments is too wide or narrow they will perform an arbitrage by selling over and buying under fair value. If, after execution costs, the trade can be entered into above or below fair value, the trader is guaranteed a profit when the prices of the two instruments converge.

Fragmentation In markets in which there is no Central Limit Order Book, such as securities, orders routed through various electronic trading platforms are often executed at prices other than the trader expected or not at all. Because ECNs and exchanges have yet to find a way for all of their respective electronic systems to communicate efficiently, the result is fragmentation.

Fundamental analysis The study of the basic economic factors that influence prices in a marketplace. Those who analyze the market based on fundamentals want to determine a market's inherent value so as to be able to capitalize on trading opportunities if, for some reason, the market pushes the price away from that value.

Fungible Interchangeable. For example, an S&P 500 futures contract is fungible with five (5) e-mini contracts.

Futures Commission Merchant (FCM) An entity that is licensed by the National Futures Association to solicit customers and broker futures trades on their behalf. FCMs are often members of the exchanges at which their customers trade.

GLOBEX The electronic trade matching platform of the CME and the partners in the GLOBEX Alliance.

GLOBEX Foreign Exchange Facility (GFX) The CME's currency market-making division. The GFX, which is comprised of traders and risk managers, provides a two-sided market for currency futures on GLOBEX, guaranteeing customers a continuous liquid market in currency futures.

Hedge Ratio The number of futures contracts needed to hedge a stock portfolio. In order to figure out the true hedge ratio, it is important to know the beta of the portfolio.

Hedging Any transaction that reduces the risk associated with holding an investment. The concept of hedging is the foundation of the futures market. By passing off an unwanted risk to a counterparty who is more willing to, or able to, take that particular risk, a hedge against the uncertainty of the future is created.

Illiquidity The absence of liquidity. In an illiquid market, it is almost impossible to determine and receive the true value of an asset.

Index arbitrage *See* Program Trading.

Introducing Broker (IB) An entity that is licensed by the National Futures Association to solicit customers and broker their trades. An IB clears the trades through a FCM.

Level II A NASDAQ electronic system that provides access to the inside market. The inside market includes the current bid and offer. Active day traders must have access to Level II in order to pursue their trading strategies.

Leverage The ability to control an investment position without having to deposit the entire purchase price immediately. The large amount of leverage available in the futures markets enables the trader to both make and lose money faster and in greater amounts (on a percentage basis) than in the equities markets.

LIFFE The London International Financial Futures Exchange. This is a completely electronic exchange whose main products are short-term European interest rates and the FTSE Index.

Liquidity The ability to buy or sell at a reasonable price whenever needed.

Local A trader at an exchange who trades using his own capital. Sometimes the phrase is used interchangeably with "market-maker," but not all locals are market-makers and not all market-makers are locals.

Margin The amount of funds that must be deposited before a security can be bought. The Federal Reserve sets margin limits for securities. In the futures markets, margins are referred to as "Performance Bonds" (PB) and are set by the exchanges and Commodity Futures Trade Commission.

Market-Maker A trader who makes a continuous two-sided market for customer orders coming into the marketplace. Market-makers rely on order flow to conduct business profitably. By receiving a premium from the customer for providing the market-making services, she has enough of an edge to make money on virtually every transaction.

Market order A buy or sell order that must be executed at the best price available at the time the order enters the marketplace.

Mark-to-the-Market The value of a futures trading position is adjusted to reflect gains and losses at the end of a trading session. Each day (in volatile periods, two or three times per day), the mark-to-the-market value is used to determine whether the account contains enough cash to maintain the minimum required performance bond to support the position.

NASDAQ An acronym for the National Association of Securities Dealer's Automated Quotation system.

National Futures Association (NFA) A self-regulating agency that enacts rules governing brokers and traders in the futures industry.

New York Mercantile Exchange (NYMEX) The premier energy futures exchange. Its main products include oil, gas, and metals. The vast majority of NYMEX business is conducted through Open Outcry. It has an electronic trade matching system called ACCESS, but one can only trade electronically when the trading floor session closes.

New York Stock Exchange (NYSE) The largest organized U.S. stock exchange. It conducts trade through a combination of electronic order routing and a floor-based specialist system. The specialists, who have access to the book of customer orders, are required to maintain an orderly market by matching buyers and sellers and taking positions into their own portfolios when there are order imbalances.

Offer The price at which one will sell an asset.

Open architecture System design based on publicly available and standardized software which can be easily linked to other systems and software.

Open Outcry The traditional methodology for conducting trade on the floors of the domestic exchanges.

Over-the-Counter (OTC) Any market that is comprised of dealers who transact business with each other in a non-regulated, non-exchange environment.

Perfect Hedge Theoretically, a hedge that perfectly offsets any gains or losses from a move in the market. A perfect hedge is rarely achievable, although it is possible to construct hedges that are almost perfect. Most hedgers are happy if the majority of the risk associated with a market exposure can be hedged.

Physical delivery When a non-cash-settled commodity futures contract expires, a short-seller has an obligation to deliver the specified amount of the commodity to a particular location (as spelled out in the contract specifications). Examples of physically delivered commodities are soybeans, gold, and live cattle.

Position limit The maximum number of futures or options contracts one can hold. The limits are set by the exchanges and the CFTC. There are also position reporting limits which require the trader to notify the CFTC if a position in a market is in excess of a specified number of contracts.

Position trader The opposite of a day trader; one who carries positions after the end of a trading session. As the markets move towards 24-hour trading, the distinction between day trader and position trader becomes somewhat blurred. Nonetheless, the position trader can be described as tending to take a longer-term view of the market's condition than does the day trader. Some position traders will carry trades for days, weeks, months, or even, in some cases, years.

Principal Any individual or entity who takes the other side of a trade. For example, if a FOREX dealer is told by his customer to buy $10 million worth of Japanese Yen and he sells the customer the Yen from his own portfolio, he has acted in the capacity of a principal.

Program trading When the price of the S&P 500 futures contract diverges from the value of the stocks in the underlying Index, there exists an arbitrage opportunity. If the futures price is undervalued relative to the stocks, then traders will buy futures and sell stocks; if futures are overvalued, they will sell futures and buy stocks.

Random Walk   The theory that price history is of no value in helping to predict the future price of a market. The foundation of the theory is that all prices—past, current, and future—simply reflect how traders react to information as it comes into the market at random.

Regulation T (Reg T)   The Federal Reserve sets margins for stock transactions. Under Reg T a trader can buy $2 worth of stock for every $1 on account. Therefore, someone with $50,000 on account could purchase up to $100,000 worth of stock on margin.

Risk management   The practice of managing one's exposure to an adverse move in the market.

Securities and Exchange Commission (SEC)   The governmental body responsible for regulating the U.S. Securities markets.

Security   A stock, bond, or an option.

SelectNet   An electronic trade matching system operated by NASDAQ in which traders can "preference," or communicate, with specific market-makers in the hope that they will be willing to trade at a particular price. Unlike a "SOES" order, which is guaranteed to be executed if there is a bid or offer available to satisfy the order, using SelectNet does not guarantee an execution. So, for instance, if a market-maker is offering to sell 5,000 shares of a stock at 15 and a trader uses SelectNet to try to buy 1,000 shares from him, the market-maker has the option of making the trade or rejecting it.

Side-by-Side trading   Some U.S. futures exchanges allow the trading of their products simultaneously in the Open Outcry pits and on the exchange's electronic platforms. Products such as Eurodollars and currencies at the CME and 30-Year Bonds and 10-Year Notes at the CBOT trade side-by-side. E-mini trading is a variation on the side-by-side approach; even though it and the large size pit-traded contract trade simultaneously and are fungible, technically it cannot be considered true side-by-side trading, because they are different products.

Single Stock Futures (SSF)   A SSF is a cash-settled futures contract on a specified number of shares of a particular stock. In early 2001, the CFTC and SEC agreed to end the long standing ban on trading single stock futures and will allow exchanges to begin trading the instruments by early 2002. A number of European exchanges got a head start on the U.S. exchanges and began trading a limited number of SSF in early 2001. At the time of this writing, they have only experienced limited success with the products. Some observers believe that SSF will be the biggest growth market ever created by the futures industry.

Sixty-Forty treatment (60/40 treatment) Futures profits are taxed at a preferential rate: 60 percent is considered long-term gain and is taxed at the 20 percent capital gains tax rate, while the remaining 40 percent is taxed as a short-term gain. The so-called 60/40 split results in a blended tax rate that is about 10 percent lower than if the profits were all taxed as short-term gains.

Slippage The amount of difference between the price at which one would expect to execute an order and where the order is actually executed. For off-the-floor customers, slippage is a significant consideration when sending orders to the pit. Because even the best price information available lags behind the trading activity in the pit, one must expect that on average there will be slippage of a certain number of points on every order. In the electronic environment, where one can view and participate in the market activity in real time, slippage becomes more manageable.

Small Order Execution System (SOES) An electronic trade matching system operated by NASDAQ that routes small orders on the system and guarantees executions when a bid and offer match.

Speculator One who is willing to assume risk to realize a trading opportunity. This is distinguished from investing in that the speculator is not particularly interested in the inherent value of an instrument, but in what she can sell it for. This is, admittedly, a fine line which often blurs. One characteristic that usually differentiates speculation from investing is that the speculator tends to look for relatively short-term trading opportunities, while the investor is focused on appreciating value over the long-term.

Spread Long one instrument and short another. Many traders watch the price relationship between different products or different expirations within a single product. When the price relationship is out of line with what the trader perceives to be the true value of the spread, he will buy the undervalued instrument and sell the one that is overvalued. If the trader's perception was correct, eventually the spread relationship will come back into line and the trader can exit the position with a profit.

Stop order   An order to buy or sell when the market touches a specified price. A stop order becomes a market order when its price is touched and is then executed at the best price available at that time. A buy stop is placed *above* the current market price and a sell stop is placed *below*. Traders, particularly technicians, tend to place stops in obvious places (e.g., above or below major trend lines). Oftentimes, floor traders will probe for the stops in order to take advantage of the customer's willingness to pay a premium for executing these types of orders.

Straight Through Processing (STP)   The capturing of trade details in the back office directly from the front-end trading system. This includes automated processing of confirmation and settlement instructions without the need for re-keying or re-formatting data.

Sydney Futures Exchange (SFE)   One of the first Open Outcry exchanges to make the transition to electronic trading. Its main products are interest rates, stock indices, and commodities. Sydney offers some interesting trading opportunities to the U.S. trader because of the time zone difference: its morning coincides with the late afternoon in the United States.

Take delivery   When one does not close out a long futures position prior to expiration she must pay for and take delivery of the commodity. Soybeans are an example of a commodity that is deliverable; each futures contract is for 5,000 bushels of beans.

Technical analysis   Also known as charting. Technical analysis is the study of market history in order to forecast future market prices.

Tick   The minimum increment a futures product must move in order to change from the prevailing price. The value of a *tic* varies from product to product and is contingent on the size of the contract. For instance, a *tic* in the Japanese Yen contract is worth $12.50, while a *tick* in the S&P 500 is worth $25.00. Markets do not, however, have to move in single *tick* increments. In volatile times traders may widen the bid/ask spread to several or many *ticks* in order to justify taking the additional risk of providing liquidity in a volatile atmosphere.

The *Tick* The difference between the number of New York Stock Exchange stocks trading at a higher price than the previous trade and the number trading at a lower price than the previous trade. In other words, the *tick* is the number of up-*tics* minus down-*tics*. Generally, as the *tick* moves beyond $+1,000$ it indicates a bullish situation; as it moves beyond $-1,000$ it indicates a bearish situation. A scenario in which the *tick* approaches 1,500 (plus or minus) is a sign that the market is moving strongly in a specific direction. More important than the absolute number of stocks higher or lower at any given moment is the relative value of the *tick* during the day, or over a series of days. Whether, for instance, the *tick* is at 200 or -200 does not tell you much about if the market is particularly strong or weak. If, however, the *tick* plunges very quickly from 200 to -200, this could be seen as an extremely bearish signal. Like all market indicators, the *tick* is subject to interpretation and every trader may construe its meaning differently. Nonetheless, it remains one of the best short-term trading indicators for those who wish to trade the S & P futures and e-mini contracts.

Two-Sided market Market-makers quote both the bid and offer, or a two-sided market. Without a two-sided market, a market is said to be illiquid.

Zero sum game Futures markets are said to be a zero sum game because for every winner there must be a loser.

APPENDIX B

Index Components: S&P 500 and NASDAQ 100

COMPONENTS OF THE NASDAQ 100 INDEX (AS OF JUNE, 2001)

3 Com Corporation
Abegenix, Inc.
ADC Telecommunications, Inc.
Adelphia Communications Corporation
Adobe Systems Incorporated
Altera Corporation
Amazon.com, Inc.
Amgen, Inc.
Apple Computer, Inc.
Applied Materials, Inc.
Applied Micro Circuits Corporation
Ariba, Inc.
AtHome Corporation
Atmel Corporation
BEA Systems, Inc.
Bed Bath & Beyond, Inc.
Biogen, Inc.
Biomet, Inc.
Broadcom Corporation
BroadVision, Inc.
Brocade Communications Systems, Inc.
Check Point Software Technologies Ltd.
Chiron Corporation
CIENA Corporation
Cintas Corporation
Citrix Systems, Inc.
Cixco Systems, Inc.
CMGI, Inc.
CNET Networks, Inc.
Comcast Corporation
Compuware Corporation
Comverse Technology, Inc.
Concord EFS, Inc.
Conexant Systems, Inc.
Costco Wholesale Corporation
Dell Computer Corporation
eBay, Inc.
EchoStar Communications Corporation
Electronic Arts, Inc.
Exodus Communications, Inc.
Fiserv, Inc.
Flextronics International Ltd.
Gemstar-TV Guide International, Inc.
Genzyme General
Human Genome Sciences, Inc.
i2 Technologies, Inc.
IDEC Pharmaceuticals Corporation
Immunex Corporation
Inktomi Corporation
Intel Corporation
Intuit, Inc.
JDS Uniphase Corporation

Juniper Networks, Inc.
KLA-Tencor Corporation
Level 3 Communications, Inc.
Linear Technology Corporation
LM Ericsson Telephone Company
Maxim Integrated Products, Inc.
McLeod USA Incorporated
MedImmune, Inc.
Mercury Interactive Corporation
Metromedia Fiber Network, Inc.
Microchip Technology Incorporated
Microsoft Corporation
Millennium Pharmaceuticals, Inc.
Molex Incorporated
Network Appliance, Inc.
Nextel Communications, Inc.
Novell, Inc.
Novellus Systems, Inc.
Oracle Corporation
PACCAR, Inc.
Palm, Inc.
PanAmSat Corporation
Parametric Technology Corporation
Paychex, Inc.
PeopleSoft, Inc.
PMC-Sierra, Inc.
Qlogic Corporation
Qualcomm Incorporate
Rational Software Corporation
RealNetworks, Inc.
RF Micro Devices, Inc.
Sanmina Corporation
Siebel Systems, Inc.
Smurfit-Stone Container Corporation
Staples, Inc.
Starbucks Corporation
Sun Microsystems, Inc.
Tellabs, Inc.
TMP Worldwide, Inc.
USA Networks, Inc.
Verisign, Inc.
VERITAS Software Corporation
Vitesse Semiconductor Corporation
Voicestream Wireless Corporation
WorldCom, Inc.
Xilinx, Inc.
XO Communications, Inc.
Yahoo, Inc.

S&P 500 COMPONENT LIST (AS OF JUNE, 2001)

Abbot Labs
Adaptec
ADC Telecomm
Adobe Systems
ADP
Advanced Micro Devices
AES Corp
Aetna
AFLAC
Agilent Technologies
AIG
Air Products & Chemicals
Alberto Culver
Albertson's
Alcoa
Allegheny Energy
Allegheny Technologies
Allergan
Allied Waste
Allstate Corp.
ALLTELL Corp.
AlteraALZA Corp.
Ambrac Financial
Amerada Hess
Ameren Corp.
American Electric
American Express
American General
American Greetings
American Home Products
American Power
Amgen
AMR Corp.
AmSouth Bancorp
Anadarko Petroleum
Analog Devices
Andrews Corp.
Anheuser-Busch
AOL Time Warner
Aon Corp.
Apache Corp.
Apple Computer

Applera Corp.
Applied Materials
Applied Micro
Archer-Daniels Midland
Ashland Inc.
AT&T
Autodesk
Autozone Inc.
Avaya
Avery Dennison
Avon Products
Baker-Hughes
Ball Corp.
Bank of America
Bank of NY
Bank One Corp.
Bard, Inc.
Barrick Gold Corp.
Bausch & Lomb
Baxter International
BB & T Corp.
Bear Stearns
Becton, Dickinson
Bed Bath & Beyond
Bell South
Bemis Company
Best Buy
Biogen
Biomet
Black & Decker
Block H & R
BMC Software
Boeing Company
Boise Cascade
Boston Scientific
Bristol-Meyer
Broadcom
Broadvision
Brown-Forman
Brunswick Corp.
Burlington Northern
Burlington Resources
Cabletron Systems

Calpine Corp.
Campbell Soup
Capital One
Cardinal Health
Carnival Corp.
Caterpillar
Cendant
Centex
Century Tel
Charles Schwab
Charter One
Chevron
Chiron
Chubb
CIGNA
Cincinnati Financial
CINergy Corp.
Cintas Corp.
Circuit City
Circuits
Cisco Systems
CIT Group
CITIGROUP
Citizen's Communications
Citrix Systems
Clear Channel
Clorox Co.
CMS Energy
Coca-Cola Co.
Coca-Cola Enterprises
Colgate-Palmolive
Comcast Class A
COMERICA
Compaq Computer
Computer Associates
Computer Scientists
Compuware Corp.
Comverse Technologies
Con-Agra Foods
Concord EFS
Convexant
Conoco, Inc.
Conseco

Consolidated Edison
Consolidated Stores
Convergys Corp.
Cooper Industries
Cooper Tire
Coors
Corning, Inc.
Costco
Countrywide Credit
Crane Company
CSX Corp.
Cummins, Inc.
CVS Corp.
Dana Corp.
Danaher Corp.
Darden Restaurants
Deere & Co.
Dell Computer
Delphi Automotive
Delta Air
Deluxe Corp.
Devon Energy
Dillard, Inc.
Dollar General
Dominion Resources
R.R. Donnelly
Dover Corp.
Dow Chemical
Dow Jones
DTE Energy
Duke Energy
Dynegy Inc.
Eastman Kodak
Eaton Corp.
Ecolab Inc.
Edison Intl.
EDS
E.I. Du Pont
El Paso Corp.
EMC Corp.
Emerson Electric
Engelhard Corp.
Enron Corp.
Entergy Corp.

EOG Resources
Equifax
Exelon
Exxon Mobil
Fannie Mae
Federal Home Loan
Federated Dept. Stores
FedEx
Fifth Third Bancorp
First Data
First Union
FirstEnergy
FiServ
Fleet Boston
Fluor Corp.
FMC Corp.
Ford Motor
Forest Laboratories
Fortune Brands
FPL Group
Franklin Resources
Freeport-McMoran
Gannet Co.
The Gap
Gateway
General Dynamics
General Electric
General Mills
General Motors
Genuine Parts
Georgia-Pacific
Gillette
Global Crossing
Golden West
Goodrich
Goodyear
GPU, Inc.
Grainger
Great Lakes Chemical
Guidant Corp.
Haliburton Corp.
Harcourt General
Harley-Davidson
Harrah's

Hartford Financial
Hasbro, Inc.
HCA
HealthSouth
Heinz
Hercules
Hershey Foods
Hewlett-Packard
Hilton Hotels
Home Depot
Homestake Mining
Honeywell
Household Intl.
Humana, Inc.
Huntington Banc
IBM
Illinois Tool Works
IMS Health
Inco Ltd.
Ingersoll-Rand
Intel Corp.
International Flavor
International Paper
Interpublic Group
Intuit Inc.
ITT Industries
Jabil Circuit
JDS Uniphase
Jefferson-Pilot
Johnson & Johnson
Johnson Controls
J.P. Morgan Chase
KB Home
Kellogg Co.
Kerr-Mcgee
Key Corp.
Key-Span
Kimberly Clark
Kinder Morgan
King Pharmaceuticals
KLA-Tencor
K-Mart
Knight Ridder
Kohl's

Kroger
Legget & Platt
Lehman Brothers
Lexmark Intl.
Lilly & Co.
The Limited
Lincoln National
Linear Technology
Liz Claiborne
Lockheed Martin
Loews Corp.
Longs Drug Store
Louisiana Pacific
Lowe's Cos.
LSI Logic
Lucent Technologies
Manor Care
Marriot Intl.
Marsh & McLennan
Masco Corp.
Mattel Corp.
Maxim Integrated Products
May Dept. Stores
Maytag Corp.
MBIA Inc.
MBNA Corp.
McDermott Intl.
McDonald's Corp.
McGraw-Hill
McKesson HBOC
Mead Corp.
MedImmune Inc.
Medtronic
Mellon Financial
Merck
Mercury Interactive
Meredith Corp.
Merrill Lynch
MetLife
MGIC
Micron Technology
Microsoft
Millipore Corp.

Mirant Corp.
Molex, Inc.
Moody's Corp.
Morgan Stanley Dean Witter
Motorola, Inc.
Nabor's Industries
National City Corp.
National Semiconductor
National Service
Navistar International
NCR Corp.
Network Appliance
New York Times
Newell Rubbermaid
Newmont Mining
NEXTEL Communications
Niagra Mohawk
NICOR, Inc.
Nike
NiSource
Noble Drilling
Nordstrom
Norfolk Southern
Nortel Networks
Northern Trust
Northrop Grunman
Novell Inc.
Novellus Systems
Nucor
Occidental Petroleum
Office Depot
Omnicom
ONEOK
Oracle Group
PACCAR, Inc.
Pactiv Corporation
Pall Corp.
Palm, Inc.
Parametric Technology
Parker-Hannifin
Paychex Inc.
Penney

People's Energy
PeopleSoft, Inc.
PepsiCo
PerkinElmer
Pfizer
PG&E Corp.
Pharmacia Corp.
Phelps Dodge
Phillip Morris
Phillips Petroleum
Pinnacle West
Pitney Bowes
Placer-Dome
PNC Financial
Potlatch
Power-One Inc.
PPG Industries
PPL Corporation
Praxair
Proctor & Gamble
Progress Energy
Progressive Corp.
Providian Financial
Public Serv. Enterprise
Pulte Corp.
QLogic Corp.
Quaker Oats
QUALCOMM Inc.
Quintiles Transnational
Qwest Communications
Radio Shack
Ralston-Purina
Raytheon
Reebok Intl.
Regions Financial
Reliant Energy
Robert Half
Rockwell Intl.
Rohm & Hass
Rowan Co.
Royal Dutch Petroleum
Ryder Systems
Sabre Holdings
SAFECO Corp.

Safeway
Sanmina Corp.
Sapient
Sara-Lee
SBC Communications
Schering-Plough
Schlumberger
Scientific Atlanta
Sealed Air
Sears Roebuck
Selectron
Sempra Energy
Sherwin Williams
Siebel Systems
Sigma-Aldrich
Snap-On
South Trust
Southern Co.
Southwest Airlines
Sprint
St. Jude Medical
St. Paul Cos.
Stanley Works
Staples
Starbucks
Starwood Hotels
State Street Corp.
Stillwell Financial
Stryker Corp.
Sun Microsystems
Sunoco, Inc.
SunTrust Banks
Supervalue
Symbol Technologies
Synovus Financial
Sysco Corp.
Target Corp.
Tektronix
Tellabs
Temple-Inland
Tenet-Healthcare
Teradyne
Texaco
Texas Instruments
Textron
Thermo Electron
Thomas and Betts
3M
Tiffany & Co.
Timkin
TJX Companies
Torchmark Corp.
Tosco Corp.
Toys Us
Transocean Sedco
Tribune Co.
TRICON Global
   Restaurant
T. Rowe Price
TRW
Tupperware
TXU
Tyco International
Unilever
Union Pacific
Union Planters
Unisys Corp.
United Health
United Technologies
Univision
   Communications
Unocal
UNUM Provident
US Airways
U.S. Bancorp
USA Education
UST Inc.
USX-Marathon
USX-U.S. Steel
Veritas Software
Verizon Communications
V.F. Group
Viacom
Visteon
Vitesse Semiconductor
Vulcan Materials
Wachovia Corp.
Walgreen
Wal-Mart
Walt Disney
Washington Mutual
Waste Management
Watson Pharmaceuticals
WellPoint Health
Wells Fargo
Wendy's Intl.
Westvaco Corp.
Weyerhauser Corp.
Whirlpool Corp.
Willamette Industries
Williams Cos.
Winn-Dixie
WorldCom
Worthington Ind.
Wrigley
Xcel Energy
Xerox
Xilinx, Inc.
Yahoo, Inc.

APPENDIX C

Contract Specifications: S&P and NASDAQ Futures Contracts

S&P 500 INDEX FUTURES

TICKER SYMBOL	SP
CONTRACT SIZE	$250 × S&P 500 Stock Index Price
MINIMUM *TICK* SIZE	.10 Index Points ($25)
	.05 Index Points ($12.50) Spreads
CONTRACT MONTHS	March, June, September, December
TRADING HOURS	Trading Floor: 8:30 a.m.-3:15 p.m.;
	GLOBEX: 3:45 p.m.-8:15 a.m.; Monday-Thursday 5:30 p.m.-8:15 a.m.; Sunday and Holidays
LAST TRADING DAY	Thursday prior to the third Friday of the contract month
POSITION LIMITS	20,000 net long or short in all contract months combined
SETTLEMENT	Cash Settlement. All open positions at the close of the final trading day are settled in cash to the Special Opening Quotation of the S&P 500 Index on Friday morning

E-MINI S&P 500 INDEX FUTURES

TICKER SYMBOL	ES
CONTRACT SIZE	$50 × S&P 500 Stock Index Price
MINIMUM *TICK* SIZE	.25 Index Points ($12.50)
	.10 Index Points ($5.00) Spreads
TRADING HOURS	Virtually 24 hour trading
CONTRACT MONTHS	March, June, September, December
LAST TRADING DAY	Trading can occur up to 8:30 a.m. (Chicago time) on the third Friday of the contract month
POSITION LIMITS	Position Limits work in conjunction with existing S&P 500 position limits

NASDAQ 100 FUTURES

TICKER SYMBOL	ND
CONTRACT SIZE	$100 × NASDAQ 100 Index Price
MINIMUM *TICK* SIZE	.50 Index Points ($50)
TRADING HOURS	Trading Floor: 8:30 a.m.-3:15 p.m.;
	GLOBEX: 3:45 p.m.-8:15 a.m. Monday-Thursday 5:30 p.m.-8:15 a.m.; Sunday and Holidays
CONTRACT MONTHS	March, June, September, December
LAST TRADING DAY	The Thursday prior to the third Friday of the contract month
POSITION LIMITS	5,000 net long or short in all contract months combined
SETTLEMENT	Cash Settlement. All open positions at the close of the final trading day are settled in cash to the Friday morning Special Opening Quotation of the NASDAQ 100 Index, computed from a five-minute volume-weighted average of each component stock's opening price.

E-MINI NASDAQ 100 INDEX

TICKER SYMBOL	NQ
CONTRACT SIZE	$20 × NASDAQ 100 Index Price
MINIMUM *TICK* SIZE	.50 Index Points ($10)
TRADING HOURS	Virtually 24 hour trading
CONTRACT MONTH	March, June, September, December
LAST TRADING DAY	Trading can occur up to 8:30 a.m. (Chicago time) on the third Friday of the contract month
POSITION LIMITS	Position Limits work in conjunction with existing NASDAQ 100 position limits

APPENDIX D
Tax Treatment of Futures

SECTION 1256 CONTRACTS MARKED-TO-THE-MARKET

If you hold a Section 1256 contract at the end of the tax year, you generally must treat it as sold at its fair market value on the last business day of the tax year.

Section 1256 Contract

A Section 1256 contract is any:

- Regulated futures contract
- Foreign currency contract
- Non-equity option
- Dealer equity option

Regulated futures contract. This is a contract that:
- Provides the amounts that must be deposited to, or can be withdrawn from, your margin account depending on daily market conditions (a system of marking to market).
- Is traded on, or subject to, the rules of a qualified board of exchange.

A qualified board of exchange is a domestic board of trade designated as a contract market by the Commodity Futures Trading Commission, any board of trade or exchange approved by the Secretary of the Treasury, or a national securities exchange registered with the Securities and Exchange Commission.

Foreign currency contract. This is a contract that:
- Requires delivery of a foreign currency that has positions traded through regulated futures contracts (or settlement of which depends on the value of that type of foreign currency).
- Is traded in the *Interbank* market.
- Is entered into at arm's length at a price determined by reference to the price in the *Interbank* market.

Bank forward contracts with maturity dates that are longer than the maturities ordinarily available for regulated futures contracts are considered to meet the definition of a foreign currency contract if the above three conditions are satisfied.

Special rules apply to certain foreign currency transactions. These transactions may result in ordinary gain or loss treatment. For details, see Internal Revenue Code, Section 988 and regulations sections 1.988-1(a)(7) and 1.988-3.

Non-equity option. This is any listed option (defined below) that is not an equity option. Non-equity options include debt options, commodity futures options, currency options, and broad-based stock index options. A broad-based stock index is based upon the value of a group of diversified stocks or securities (such as the Standard and Poor's 500 Index).

Warrants based on a stock index that are economically and substantially identical in all material respects to options based on a stock index are treated as options based on a stock index.

Cash-settled options. Cash-settled options based on a stock index, which are either traded on or subject to the rules of a qualified board of exchange, are non-equity options if the Securities and Exchange Commission (SEC) determines that the stock index is broad based.

Listed option. This is any option that is traded on, or subject to the rules of a qualified board or exchange (as discussed earlier under Regulated futures contracts). A listed option, however, does not include an option that is a right to acquire stock from the issuer.

Dealer equity option. This is any listed option that, for an options dealer:
- Is an equity option

- Is bought or granted by that dealer in the normal course of the dealer's business activity of dealing in options
- Is listed on the qualified board of exchange where the dealer is registered

An *options dealer* is any person registered with an appropriate national securities exchange as a market-maker or specialist in listed options.

Equity Option. This is any option:
- To buy or sell stock
- Valued directly or indirectly by reference to any stock, group of stocks, or stock index.

Equity options include options on certain narrow-based stock indices but exclude options on broad-based stock indices and options on stock index futures.

An equity option, however, does not include an option for any group of stocks or stock index if:

- The Commodities Future Trading Commission has designated a contract market for a contract based on that group or index and that designation is in effect
- The Secretary of the Treasury determines that the option meets the legal requirements for such a designation.

Marked-to-the-Market Rules

A Section 1256 contract that you hold at the end of the tax year will generally be treated as sold at its fair market value on the last business day of the tax year, and you must recognize any gain or loss that results. That gain or loss is taken into account in figuring your gain or loss when you later dispose of the contract, as shown below in the example under the 60/40 rule.

Hedging exception. The marked-to-the-market rules do not apply to hedging transactions. See Hedging Transactions, which appears later.

60/40 rule. Under the marked-to-the-market system, 60 percent of your capital gain or loss will be treated as a long-term capital gain or loss and 40 percent will be treated as a short-term capital gain or loss. This is true regardless of how long you actually held the property.

Example. On June 23, 1999, you bought a regulated futures contract for $50,000. On December 31, 1999, (the last business day of the tax year) the fair market value of the contract was $57,000. You have a $7,000 gain on your 1999 tax return treated as 60 percent long-term and 40 percent short-term capital gain.

On February 2, 2000, you sold the contract for $56,000. Because you already recognized a $7,000 gain on your 1999 return, you recognize a $1,000 loss ($57,000-$56,000) on your 2000 tax return, treated as a 60 percent long-term and 40 percent short-term capital loss.

Limited partners or entrepreneurs. The 60/40 rule does not apply to dealer equity options that result in capital gain or loss allocatable to limited partners or limited entrepreneurs (defined later under Hedging Transactions). Instead, these persons should treat all these gains or losses as short-term under the marked-to-the-market system.

Terminations and transfers. The marked-to-the-market rules also apply if your obligation or rights under Section 1256 contracts are terminated or transferred during the tax year. In this case, use the fair market value of each Section 1256 contract at the time of termination or transfer to determine the gain or loss. Terminations or transfers may result from any offsetting, delivery, exercise, assignment, or lapse of your obligation or rights under Section 1256 contracts.

Loss carryback election. An individual or partnership having a net Section 1256 contracts loss can elect to carry this loss back three years instead of carrying it over to the next year. See *How to Report* for information about reporting this election on your return.

The loss carried back to any year under this election cannot be more than the net Section 1256 contracts gain in that year. In addition, the amount of loss carried back to an earlier tax year cannot increase or produce a net operating loss for that year.

The loss is carried to the earliest carryback year first, and any unabsorbed loss amount can then be carried to each of the next two tax years. In each carryback year, treat 60 percent of the carryback amount as a long-term capital loss and 40 percent as a short-term capital loss from Section 1256 contracts.

Do not treat any part of a net Section 1256 contracts loss carried to 1997 as a 28 percent rate gain or loss. For 1997, certain long-term capital gains and losses from property held 18 months or less were treated as 28 percent rate gains and losses (discussed under *Reporting Capital Gains and Losses*).

If only a portion of the net Section 1256 contracts loss is absorbed by carrying the loss back, the unabsorbed portion can be carried forward, under the capital loss carryover rules to the year following the loss (see *Capital Losses under Reporting Capital Gains and Losses*). Figure your capital loss carryover: for the loss year you had an additional short-term capital gain of 40 percent of the amount of the net Section 1256 contracts loss absorbed in the carryback years and an additional long-term capital gain of 60 percent of the absorbed loss. In the carryover year, treat any capital loss carryover from losses on Section 1256 contracts for that year.

Net Section 1256 contracts loss. This loss is the lesser of:
- The net capital loss for your tax year determined by taking into account only the gains and losses from Section 1256 contracts.
- The capital loss carryover to the next tax year determined without this election.

Net Section 1256 contracts gain. This gain is the lesser of:
- The capital gain net income for the carryback year determined by taking into account only gains and losses from Section 1256 contracts.
- The capital gain net income for that year.

Figure your net Section 1256 contracts gain for any carryback year without regard to the net Section 1256 contracts loss for the loss year or any later tax year.

TRADERS IN SECTION 1256 CONTRACTS

Gain or loss from the trading of Section 1256 contracts is capital gain or loss subject to the marked-to-the-market rules. However, this does not apply to contracts held to the purposes of hedging property if any loss from the property would be an ordinary loss.

Treatment of underlying property

The determination of whether an individual's gain or loss from any property is ordinary or capital gain or loss is made without any regard to the fact that the individual is actively engaged in dealing in or trading Section 1256 contracts related to that property.

How to Report

If you disposed of regulated futures or foreign currency contracts in 2000 (or had unrealized profit or loss on these contracts that were open at the end of 1999 or 2000), you should receive Form 1099-B or an equivalent statement, from your broker.

Form 6781. Use Part I of Form 6781, *Gains and Losses From Section 1256 Contracts and Straddles,* to report your gains and losses from all Section 1256 contracts that are open at the end of the year or that were closed out during the year. This includes the amount shown in Box 9 of Form 1099-B. Then enter the net amount of these gains and losses on Schedule D (Form 1040). Include a copy of Form 6781 with your Income Tax Return.

If the Form 1099-B you receive includes a straddle or hedging transaction, defined later, it may be necessary to show certain adjustments on Form 6781. Follow the Form 6781 instructions for completing Part I.

Loss carryback election. To carryback the loss under the election procedures described earlier, file Form 1040X or an appropriate amended return for the year to which you are carrying the loss with an amended Form 6781 attached. Follow the instructions for completing Form 6781 for the loss year to make this election.

Hedging Transactions

The marked-to-the-market rules, described earlier, do not apply to hedging transactions. A transaction is a hedging transaction if both of the following conditions are met:

1. You entered into the transaction in the normal course of your trade or business primarily to manage the risk of:

a) Price changes or currency fluctuations on ordinary property you hold (or will hold)

b) Interest rate, or price changes, or currency fluctuations, on your current or future borrowings or ordinary obligations

2. You clearly identified the transaction as being a hedging transaction before the close of the day on which you entered into it.

This hedging transaction exception does not apply to transactions entered into by or for any syndicate. A syndicate is a partnership, corporation, or other entity (other than a regular corporation) that allocates more than 35 percent of its losses to limited partners or limited entrepreneurs. A limited entrepreneur is a person who has an interest in an enterprise (but not as a limited partner) and who does not actively participate in its management. However, an interest is not considered held by a limited partner or entrepreneur if the interest holder actively participates (or did so for at least five full years) in the management of the entity or is the spouse, child (including a legally adopted child), grandchild, or parent of an individual who actively participates in the management of the entity.

Hedging loss limit. If you are a limited partner or entrepreneur in a syndicate, the amount of a hedging loss you can claim is limited. A "hedging loss" is the amount by which the allowable deductions in a tax year that resulted from a hedging transaction (determined without regard to the limit) are more than the income received or accrued during the tax year from this transaction.

Any hedging loss that is allocated to you for the tax year is limited to your taxable income for that year from the trade or business in which the hedging transaction occurred. Ignore any hedging transaction items in determining this taxable income. If you have a hedging loss that is disallowed because of this limit, you can carry it over to the next tax year as a deduction resulting from a hedging transaction.

If the hedging transaction relates to property other than stock or securities, the limit on hedging losses applies if the limited partner or entrepreneur is an individual.

The limit on hedging losses does not apply to any hedging loss to the extent that it is more than all your unrecognized gains from hedging transactions at the end of the tax year that are from the trade or business in which the hedging transaction occurred.

Sale of property used in a hedge. Once you identify personal property as being part of a hedging transaction, you must treat gain from its sale or exchange as ordinary income, not capital gain.

APPENDIX E
The Investor's Bill of Rights

MAKING INFORMED DECISIONS

In many important ways, an investor is not simply a consumer but a party to a legal contract. Both the offeror and purchaser of an investment have rights and responsibilities. This "Bill of Rights" is designed to assist you, the investor, in making an informed decision before committing your funds. It is not intended to be exhaustive in its descriptions. Should you desire further information about a particular type of investment, you are invited to contact the appropriate organization listed at the end of this brochure.

HONESTY IN ADVERTISING

Many individuals first learn of investment opportunities through advertising in a newspaper or magazine, on radio, television, the Internet, or by mail. Phone solicitations are also regarded as a form of advertising. In practically every area of investment activity, false or misleading advertising is against the law and subject to civil, criminal, or regulatory penalties.

Bear in mind that advertising is able to convey only limited information, and the most attractive features are likely to be highlighted. Accordingly, it is never wise to invest solely on the basis of an advertisement. The only bona fide purposes of advertising are to call your attention to an offering and encourage you to obtain additional information.

FULL AND ACCURATE INFORMATION

Before you make an investment, you have the right to seek and obtain information about the investment. This includes information that accurately conveys all the material facts about the investment, including the major factors likely to affect its performance.

You also have the right to request information about the firm or the individuals with whom you would be doing business and whether they have a "track record." If so, you have the right to know what it has been and whether it is real or "hypothetical." If they have been in trouble with regulatory authorities, you have the right to know this. If a rate of return is advertised, you have the right to know how it is calculated and any assumptions it is based on. You also have the right to ask what financial interest the seller of the investment has in the sale.

Ask for all available literature about the investment. If there is a prospectus, obtain it and read it. This is where the bad as well as the good about the investment has to be discussed. If an investment involves a company whose stock is publicly traded, get a copy of its latest annual report. It can also be worthwhile to check out the Internet or visit your public library to find out what may have been written about the investment in recent business or financial periodicals.

Obtaining information is not likely to tell you whether or not a given investment will be profitable, but what you are able to find out—or unable to find out—could help you decide if it is an appropriate investment for you at that time. No investment is right for everyone.

DISCLOSURE OF RISKS

Every investment involves some risk. You have the right to find out what these risks are prior to making an investment. Some, of course, are obvious: shares of stock may decline in price, a business venture may fail, or an oil well may turn out to be a dry hole. Others may be less obvious. Many people do not fully understand that even a U.S. Treasury Bond may fluctuate in market value prior to maturity, or that with some investments it is possible to lose more than the amount initially invested. The point is that different investments involve different kinds of risk and these risks can differ in degree. A general rule of thumb is that the greater the potential reward, the greater the potential risk.

In some areas of investment, there is a legal obligation to disclose the risks in writing. If the investment does not require a prospectus or written risk disclosure statement, you might nonetheless want to ask for a written explanation of the risks. The bottom line is that unless your understanding of the ways you can lose money is equal to your understanding of the ways you can make money, do not invest!

EXPLANATION OF OBLIGATIONS AND COSTS

You have the right to know, in advance, what obligations and costs are involved in a given investment. For instance, does the investment involve a requirement that you must take some specific action by a particular time? Or is there a possibility that at some future time or under certain circumstances you may be obligated to come up with additional money?

Similarly, you have the right to a full disclosure of the costs that will be, or may be, incurred. In addition to commissions, sales charges, or "loads" when you buy and/or sell, this includes any other transaction expenses, maintenance or service charges, profit-sharing arrangement, redemption fees or penalties, and the like.

TIME TO CONSIDER

You earned the money and you have the right to decide for yourself how you want to invest it. That includes sufficient time to make an informed and well-considered decision. High pressure sales tactics violate the spirit of the law, and most investment professionals will not push you into making uninformed decisions. Thus, any such efforts should be grounds for suspicion. An investment that "absolutely has to be made right now" probably should not be made at all.

RESPONSIBLE ADVICE

Investors enjoy a wide range of different investments to choose from. Taking into consideration your financial situation, needs, and investment objectives, some are likely to be suitable for you and others are not, perhaps because of risks involved and perhaps for other reasons. If you rely on an investment professional for advice, you have the right to responsible advice.

In the securities industry, for example, "suitability" rules require that investment advice be appropriate for the particular customer. In the commodity futures industry a "know your customer" rule requires that firms and brokers obtain sufficient information to assure that investors are adequately informed of the risks involved. Beware of someone who insists that a particular investment is "right" for you although he or she knows nothing about you.

BEST EFFORT MANAGEMENT

Every firm and individual that accepts investment funds from the public has the ethical and legal obligation to manage money responsibly. As an investor, you have the right to expect nothing less.

Unfortunately, in any area of investment, there are those few less-than-ethical persons who may lose sight of their obligations, and of your rights, by making investments you have not authorized, by making an excessive number of investments for the purpose of creating additional commission income for themselves, or, at the extreme, by appropriating your funds for their personal use. If there is even a hint of such activities, insist on an immediate and full explanation. Unless you are completely satisfied with the answer, ask the appropriate regulatory or legal authorities to look into it. It is your right.

COMPLETE AND TRUTHFUL ACCOUNTING

Investing your money should not mean losing touch with your money. It is your right to know where your money is and the current status and value of your account. If there have been profits or losses, you have the right to know the amount and how and when they were realized or incurred. This right includes knowing the amount and nature of any and all charges against your account.

Most firms prepare and mail periodic account statements, generally monthly. And you can usually obtain interim information on request. Whatever the method of accounting, you have both the right to obtain this information and the right to expect that it be timely and accurate.

ACCESS TO YOUR FUNDS

Some investments include restrictions as to whether, when, or how you can have access to your funds. You have the right to be clearly informed of any restrictions in advance of making the investment. Similarly, if the investment may be illiquid—difficult to quickly convert to cash—you have the right to know this beforehand. In the absence of restrictions or limitations, it is your money and you should be able to have access to it within a reasonable period of time.

You should also have access to the person or firm that has your funds. Investment scam artists are well versed in ways of finding you but, particularly once they have your money in hand, they can make it difficult or impossible for you to find them.

RECOURSE, IF NECESSARY

Your rights as an investor include the right to seek an appropriate remedy if you believe someone has dealt with you—or handled your investment—unfairly or dishonestly. Indeed, even in the case of reasonable misunderstandings, there should be some way to reconcile differences.

It is wise to determine before you invest what avenues of recourse are available to you if they should be needed. One means of exercising your right of recourse may be to file suit in a court of law. Or you may be able to initiate arbitration, mediation, or reparation proceedings through an exchange or a regulatory organization.

APPENDIX F
Using the World Wide Web

EXCHANGES AND TRADING WEBSITES

American Stock Exchange: *www.amex.com*
Bolsa de Mercadorias & Futures (BM & F): *www.bmf.com.br*
Bourse de Montreal: *www.me.org*
Cantor Exchange: *http://cx.cantor.com*
Chicago Board Options Exchange: *www.cboe.com*
Chicago Board of Trade: *www.cbot.com*
Chicago Mercantile Exchange: *www.cme.com*
eSpeed: *www.espeed.com*
EUREX: *www.eurexchange.com*
EuroNext Paris: *www.euronext.com*
FOREX Capital Markets (FXCM): *www.fxcm.com*
FXAll: *www.e-globallink.com*
Hong Kong Exchanges (HKEX): *www.hkex.com.hk*
Intercontinental Exchange: *www.intcx.com*
International Petroleum Exchange: *www.ipe.uk.com*
International Securities Exchange: *www.iseoptions.com*
Kansas City Board of Trade: *www.kcbt.com*
London International Financial Futures Exchange (LIFFE): *www.liffe.com*
MEFF Renta Fija (MEFF): *www.meff.es*
New York Board of Trade: *www.nybot.com*

New York Mercantile Exchange: *www.nymex.com*
Osaka Securities Exchange: *www.ose.or.jp*
Pacific Exchange: *www.pacificex.com*
Singapore Exchange (SGX): *www.singaporeexchange.com*
Sydney Futures Exchange: *www.sfe.com.au*
Tokyo Commodities Exchange: *www.tocom.or.jp*
Tokyo International Financial Futures Exchange: *www.tiffe.or.jp*

BOOK VENDORS, PUBLICATIONS, AND INSTRUCTIONAL VIDEOS

Active Trader Magazine: *www.activetradermag.com*
Commodity Price Charts: *www.pricecharts.com*
Derivatives Week: *www.derivativesweek.com*
Futures and Options World Magazine: *www.fow.com*
Futures Magazine: *www.futuresmag.com*
John Wiley & Sons Inc.: *www.wiley.com*
Technical Analysis of Stocks & Commodities Magazine: *www.traders.com*
Trader's Library: *www.traderslibrary.com*
Trader's Press: *www.traderspress.com*
Trader's World Magazine: *www.tradersworld.com*
Windsor Books: *www.windsorpublishing.com*

TECHNICAL ANALYSIS PACKAGES

Aspen Graphics: *www.aspenres.com*
Commodity Quote Graphics: *www.cqg.com*
Futures Source/Bridge: *www.futuresource.com*
JS Services: *www.jsservices.com*
Local Knowledge: *www.localknowledge.com*
Meta Stock: *www.metastock.com*
Omni Trader: *www.omnitrader.com*
Trade Station Technologies: *www.tradestationtechnologies.com*

NEWS AND INFORMATION

Applied Derivatives: *www.appliedderivatives.com*
Bloomberg: *www.bloomberg.com*
Bridge: *www.bridge.com*
CBS Marketwatch: *www.cbs.marketwatch.com*
CNBC: *www.cnbc.com*
CNN: *www.cnnfn.com*
Eurozone Advisors: *www.eurozoneadvisors.com*
Financial Times: *www.ft.com*
Investors Alley: *www.investorsalley.com*
The Motley Fool: *www.fool.com*
Reuters Money Net: *www.moneynet.com*
The Street.com: *www.thestreet.com*
Zacks Investment Services: *www.zacks.com*

GOVERNMENT AND NOT-FOR-PROFIT ORGANIZATIONS

American Association of Individual Investors: *www.aaii.com*
Commodity Futures Trading Commission: *www.cftc.gov*
Edgar (SEC filings): *www.edgar-online.com*
Futures and Options Association (United Kingdom): *www.foa.co.uk*
Futures Industry Institute: *www.fiafii.org*
Gloriamundi (Articles on risk management): *www.gloriamundi.org*
International Swaps and Derivatives Association: *www.isda.org*
Managed Futures Association: *www.mfainfo.org*
National Association of Securities Dealers: *www.nasd.com*
National Futures Association: *www.nfa.org*
Securities and Exchange Commission: *www.sec.gov*

APPENDIX G
Suggested Reading

Achelis, Steven. **Technical Analysis: From A to Z,** McGraw-Hill, 2001. *A comprehensive guide to various technical approaches.*

Anuff, Joey, and Gary Wolf. **Dumb Money: Adventures of a Day Trader,** Random House, 2000. *An accurate, and often hilarious, account of what it's like to day trade NASDAQ stocks online.*

Dalton, James, F., Eric T. Jones, and Robert B. Dalton. **Mind Over Markets: Power Trading With Market Generated Information,** Probus Publishing, 1990. *Everything you ever wanted to know about the technical approach known as "Market Profile."*

Edwards, Robert D., and John Magee. **Technical Analysis of Stock Trends,** Sixth Edition, New York Institute of Finance, 1992. *This book is a classic text. First published in 1948, it forms the basis for much of the recent work performed by modern-day technicians.*

Friedfertig, Marc, and George West. **The Electronic Day Trader,** McGraw-Hill, 1998. *One of the first books available on NASDAQ Level II trading.*

Friedfertig, Marc, George West, and Jonathan Burton. **Electronic Day Trader's Secrets,** McGraw-Hill, 1999. *Friedfertig, West and Burton interview the top traders at their firm, Broadway Trading. Fascinating insights from some of the most active online securities traders in the world.*

Graham, Benjamin, David L. Dodd, and Sidney Cottle. **Security Analysis: Principles and Technique,** Fourth Edition, McGraw-Hill, 1962. *This classic text on value investing and fundamental analysis, first published in 1934, is still a valuable resource today.*

Krutsinger, Joe. **The Trading Systems Tool Kit,** Probus Publishing, 1994. *A good starter book for anyone who wishes to learn how to write and test trading models.*

Malkiel, Burton. **A Random Walk Down Wall Street,** Sixth Edition, W.W. Norton, 1999. *Another classic text. Malkiel tries to demonstrate that the market is impossible to predict because it is a "random walk."*

Miller, Merton H. **Financial Innovations & Market Volatility,** Blackwell Press, 1991. *One of the best primers I have read on how the derivatives markets work.*

Murphy, John J. **Technical Analysis of Financial Markets,** New York Institute of Finance, 1999. *Another great text covering a wide variety of technical approaches.*

Niederhoffer, Victor. **The Education of a Speculator,** John Wiley & Sons, 1997. *Niederhoffer comments on the markets. It is particularly notable and cautionary because, shortly after this book was published, Niederhoffer blew out his hedge fund with an over-leveraged stock index trade.*

Pardo, Robert. **Design, Testing, and Optimization of Trading Systems,** John Wiley & Sons, 1992. *Excellent high-level analysis of how to develop and use profitable trading models.*

Pring, Martin. **Technical Analysis Explained**, Third Edition, McGraw-Hill, 1991. *Another excellent text covering a wide array of technical approaches.*

Reuters Financial Trading Series. **An Introduction to the Bond Markets,** John Wiley & Sons, 1999. *Great primer for anyone who wants to understand the inner workings of the credit markets.*

Reuters Financial Trading Series. **An Introduction to the Derivatives Markets,** John Wiley & Sons, 1999. *A great primer for anyone who wants to understand the inner workings of the derivatives markets.*

Schwager, Jack D. **Fundamental Analysis,** John Wiley & Sons, 1995. *How to use fundamental analysis in the real world.*

Schwager, Jack D. **Market Wizards,** Prentice Hall Trade, 1989. *Classic interviews with some of the best traders in the world. Sets the standard for the "trader interview" genre.*

Schwager, Jack D. **The New Market Wizards,** Harper Business, 1994. *A follow-up to Market Wizards. Equally engrossing.*

Schwager, Jack D. **Stock Market Wizards,** Harper Business, 2001. *Interviews with some of the world's most proficient equities traders. As fascinating as the first two Wizard books.*

Schwager, Jack D. **Technical Analysis,** John Wiley & Sons, 1995. *The companion volume to Fundamental Analysis. The two books together cover just about every conceivable fundamental and technical approach to trading.*

Soros, George. **The Alchemy of Finance,** John Wiley & Sons, 1994. *The legendary money manager's manifesto on trading and economics. Great title. Second only to Charles MacKay's classic, Extraordinary Popular Delusions and the Madness of Crowds.*

Train, John. **The Money Masters,** Harper Business, 1980. *Profiles of great money managers.*

Train, John. **The New Money Masters,** Harper Business, 1989. *Follow-up to The Money Masters. Same concept, different money managers.*

Zipf, Robert. **How the Bond Market Works,** New York Institute of Finance, 1997. *Another good primer for anyone who wants to understand the basics of the bond market.*

Index

ACCESS (NYMEX), 54
Active Trader magazine, 195, 265
Agricultural futures, 98
American Association for Individual Investors, 266
American Stock Exchange (AMEX), 264
Applied Derivatives, 265
Applied Programming Interface (API), 44, 117, 120
Arbitrage, 144–146, 231
Aspen Graphics, 265
Aspire Trading Company, 221

Bank of International Settlements, 161
Basis, 178–179
Battery Ventures, 113
Beattie, Pace, 32–46
Beta, 232
Bid/Ask spread, 202–203
Blackstone Group, 113
Bloomberg, 266
Bolsa de Mercadorias & Futuros (BM&F), 92, 264
Bouroudjian, Jack, 23–32
Bourse de Montreal, see *Montreal Exchange*
Broadway Trading Company, 19–20
Brumfield, Harris, 121

B2B Marketplace (CME), 93, 115
Bund futures, 84

Cantor Fitzgerald, 114–115, 264
Cap Gemini Ernst & Young, 113
Carry, 178–179
Cattle futures, 91
CBS Marketwatch, 266
Central Limit Order Book (CLOB), 13–15, 232
Chicago Board of Trade (CBOT), 3, 35, 54, 77, 101–105, 112, 118, 121, 126, 131, 142, 155, 173–190, 232, 264
Chicago Board Options Exchange (CBOE), 27, 48, 155, 264
Chicago Mercantile Exchange (CME), 1–5, 78–79, 89–100, 118, 121, 126, 131, 149–150, 155, 159–160, 232, 264
Circuit breakers, 26, 82–83
Clearing fees
  CBOT, 105–106
  CME, 100, 148–150
  EUREX, 112
  LIFFE, 114
Clearinghouse, 93–94, 233
CNBC, 16, 31, 266
CNN, 266
Commerzbank, 30

INDEX

Commerz Futures, 23–24
Commissions (for e-minis), 150–151, 225
Commodity Futures Modernization Act (CFMA), 27–28, 155–157
Commodity Futures Trading Commission (CFTC), 27–28, 156, 162, 233, 266
Commodity Price Charts magazine, 265
Commodity Quote Graphics (CQG), 265
Connectivity, 36–44, 225–226, 233
Consumer Price Index (CPI), 188–189
Cramer, Jim, 220
Crossfire, 123–125
Currency cash market transactions, 164–169
Currency futures, 78–79, 98, 165–171, 229

Day trading, 51, 233
Decimalization, 153, 233
de Kwiatkowsi, Henryk, 206–207
Demutalization, 83–87, 92, 233
Derivatives Week, 265
Direct access, 9–13, 36–37, 50, 101, 233
Discount rate, 186
Doppelt, Paul, 46–53
Durable Goods Orders, 189

EasyScreen, 126–127
EdgarOnline, 266
Edwards and Magee, 218
Electronic Broking System (EBS), 160–161
Electronic Communication Networks (ECNs), 13–15, 130, 151, 157, 234
E-minis, 24–25, 97, 141–155, 249–250
eSpeed Exchange, 114–115, 264

EUREX, 3–4, 54, 84, 101, 110–112, 121, 126, 131, 141, 234, 264
Eurodollar futures, 79, 81, 92
EuroNext Paris, 264
Eurozone Advisors, 266
Exchange membership privileges, 146–148

Fair Value, 136–140, 234
Federal Reserve, 173, 185–190
Financial Times, 266
FOREX Capital Markets (FXCM), 264
FOREX Trading, 159–171
Fragmentation, 13–15, 234
Friedman, Milton, 78–79, 161
Front-end software, 117–132
Future Dynamics, 123–125
Futures Commission Merchant (FCM), 118, 234
Futures exchanges, 77–79, 83–87
Futures Industry Institute, 266
Futures magazine, 265
Futures and Options Association, 266
Futures and Options World magazine, 265
Futures Source/Bridge, 265
FXAll, 264

GL Trade, 128–129
Globex, 54–72, 92, 94–98, 131, 234
Globex Foreign Exchange Facility (GFX), 170–171, 234
Gloriamundi, 266
Gross Domestic Product (GDP), 189

Hedging, 74–76, 134–136, 180–185, 235, 237
Hong Kong Exchanges, 264
Housing Starts, 189

Illiquidity, 81–83, 234

INDEX

Independent Software Vendors (ISVs), 41–45, 102, 112–113, 117–132
Index Arbitrage (Program Trading), 138–140, 235
Index futures, 98
Industrial Production, 189
Instinet, 13
Interactive Brokers, 130–131
Interbank market, 161–168
Intercontinental Exchange (ICE), 264
International Petroleum Exchange (IPE), 264
International Securities Exchange (ISE), 264
International Swaps and Derivatives Association (ISDA), 266
Internet, 39–43
Internet Service Provider (ISP), 42–43
Investor's Alley, 266
Investor's Bill of Rights, 208–209, 259–263
Island, 13

John Wiley & Sons, 265
JS Services, 265

Kansas City Board of Trade, 264

Leading Indicators, 189
LIFFECONNECT, 54, 112–114
Liquidity, 76–83, 235
Local Knowledge, 265
London International Financial Futures Exchange (LIFFE), 3–4, 54, 84, 93, 126, 131, 141, 155, 235, 264

Malkiel, Bernard, 211–220
Managed Futures Association (MFA), 266
Marche A Terme d'Instruments Financiers (MATIF), 15, 141

Market making, 76, 80–81, 146–148, 236
Meff Renta Fija (MEFF), 92, 264
Melamed, Leo, 78–79
Meta Stock, 265
Montreal Exchange, 92, 264
Motley Fool, The, 266

NASDAQ Level II, 13–15, 147, 235
NASDAQ 100 Index, 141, 143, 242–244
National Association of Securities Dealers (NASDAQ), 12–15, 16, 130, 151, 236, 266
National Futures Association, 208, 236, 266
Network construction, 17–18, 32–46
New York Board of Trade, 265
New York Mercantile Exchange (NYMEX), 54, 155, 236, 265
New York Stock Exchange (NYSE), 82, 130, 236

Omni Trader, 265
Opening range, 197
Open outcry, 79, 84–87, 237
Osaka Securities Exchange, 265

Pacific Exchange, 265
Paris Bourse, 92, 113
patsystems, 129–130
Persico, Mike, 32–46
Personal Computer (choice of), 45–46, 226–227
Personal Income, 189
Pork Belly futures, 1, 81, 91
Position size, 198–199
Preuss, Andreas, 121
Producer price Index (PPI), 189
Program Trading (Index Arbitrage), 138–140, 237

Random Walk theory, 211–220, 237
Real Time Systems (RTS), 128

Redibook, 13
Regulation T (Reg T), 151, 237
Reserve Requirements, 186–187
Reuters, 160–161
Reuter's Money Net, 266
Risk management, 51, 74–76, 93, 203–206, 237
Risk/Reward ratio, 198, 224

Securities and Exchange Commission (SEC), 27–28, 156, 237, 266
Selectnet, 13, 151
Short selling, 51, 152
Singapore Stock Exchange (SGX), 92–93, 126, 265
Single Stock futures, 27–28, 53, 114, 155–157, 229, 237
Sklarew, Arthur, 214
Small Order Execution System (SOES), 13, 151, 239
Speculators, 76–77, 239
Standard & Poor 500 futures, 79, 134–140, 248
Standard & Poor 500 index, 244–247
Stock Index futures, 92, 133–157
Swiss Options and Futures Exchange (SOFFEX), 110
SYCOM, 54, 109
Sydney Futures Exchange (SFE), 54, 106–110, 126, 141, 240, 265
System-based trading, 195–196, 201–202

Tax Treatment (of futures transactions), 153–155, 239, 251–258
Technical analysis, 211–220, 229–230, 240

Technical Analysis of Stocks and Commodities magazine, 265
The Street.com, 266
Tick indicator, 140, 241
Tokyo Commodities Exchange, 265
Tokyo International Financial Futures Exchange (TIFFE), 265
Tokyo Stock Exchange, 93
Trade entry, 193–194
Trader's Library, 265
Trader's Press, 265
Trader's World magazine, 265
Trade Station Technologies, 265
Trading Arcade, 15–21, 231
Trading Floor, 1–5, 9–11, 29–31, 35–36, 47–48, 52, 144–146
Trading Losses, 51, 209–211, 224
Trading methodology, 196–199
Trading Products
  CBOT, 102–103
  CME, 98–99
  EUREX, 112
  LIFFE, 114
Trading systems, 195–196, 201–202
Trading Technologies, 121–123
Treasury Bond futures, 34, 79, 173–190
Treasury Note futures, 173–190

Unemployment report, 189

Volume, 197

Windsor Books, 265

YesTrader, 131–132

Zack's Investment Service, 266